Why Elephants Cry

Why Elephants Cry is a fascinating frolic through the literature and evidence surrounding the use of unusual behaviour of animals to measure and predict the environment. The role of animals, from the smallest ant to the biggest elephant, as predictors of environmental changes is framed around the climate crisis, which highlights the increasingly important part that animals will have to play in the future.

Renowned biologist Professor John T. Hancock collects anecdotal stories and myths along with scientific evidence, demonstrating that observation of animals can be of tangible use. He looks at the measurement of the air temperature using ants, crickets and snakes, and goes on to assess the evidence that the observation of a wide range of animals can predict the weather or the imminent eruption of volcanoes and earthquakes. Evidence of animals being able to predict lunar and solar events, such as lunar cycles and the Northern Lights, is also considered.

This is the only time that all this literature has been brought together in one place, a fascinating reference for anybody interested in animals and the environment. The book is also an ideal supplementary textbook for students studying animal behaviour.

Why Elephants Cry

How Observing Unusual Animal Behaviours Can Predict the Weather (and Other Environmental Phenomena)

John T. Hancock

CRC Press
Taylor & Francis Group
Boca Raton London New York

CRC Press is an imprint of the
Taylor & Francis Group, an **informa** business

First edition published 2023
by CRC Press
6000 Broken Sound Parkway NW, Suite 300, Boca Raton, FL 33487–2742

and by CRC Press
4 Park Square, Milton Park, Abingdon, Oxon, OX14 4RN

CRC Press is an imprint of Taylor & Francis Group, LLC

ISBN: 978-1-032-38179-4 (hbk)
ISBN: 978-1-032-38176-3 (pbk)
ISBN: 978-1-003-34384-4 (ebk)

DOI: 10.1201/9781003343844

Typeset in Bembo
by Apex CoVantage, LLC

Dedication

To my family,
past and present.

"Don't knock the weather; nine-tenths of the people couldn't start a conversation if it didn't change once in a while." ~ Kin Hubbard

Contents

Preface

Most humans can sense their surroundings. Eyes allow us to see, ears to hear. We can feel the wind on our skin and sense the temperature. We can smell if food has become rotten or is fresh to eat. We can become aware of danger. All these things allow us to survive: to live to reproduce and allow the species to continue. Humans have also developed complex tools to aid in measuring our surroundings. We have thermometers to measure the temperature, barometers to assess the air pressure and seismometers to tell us the strength of earthquakes.

All animals also need to be able to sense their environment. This idea can be extended down to the cellular level. All cells need to be able to respond to their surroundings, and often the response is altered depending on the temperature, or the presence of toxins or hormones. Cells contain multiple receptors which transmit messages into the cell, allowing the correct response to be mounted. This response may be altered metabolism or the production of new proteins. Therefore, from single-celled organisms to humans, there is a need to sense what is on the outside and change activities accordingly.

Animals often need to survive in a harsh environment. Unlike many humans, a wild animal can't usually pop down to the local supermarket or take-away – although such establishments are good scavenge sites for rats and squirrels, and even bears. They cannot shrug on a coat or slip one off. Therefore, animals will monitor the environment and have altered activities in response to changing conditions. Animals migrate and seek warmer climes or hibernate to survive the winter. If the activities of animals could be understood, would this allow humans to either measure or predict environmental conditions? This book attempts to bring together examples from the animal kingdom where this has been proposed, although it is not intended as an exhaustive list. The evidence draws from a variety of sources, including peer-reviewed journals, as well as the internet where it is much more difficult to establish any measure of robustness for the information given. Nevertheless, some anecdotal evidence is used here, and then other sources sought to establish how strong the evidence may be. There is no date cut-off used either, with some of the literature going back well into the 19th Century, or beyond.

What struck me on doing this research is, firstly, that the idea of observing animals for assessing our environment is certainly not a novel idea. Secondly, there were some extremely famous people who seemed to have dabbled in this topic as a side-line whilst they were also amazingly successful in other fields of endeavour. Often it is hard to see what attracted them to such investigations, except for mere curiosity. To all those people through history, I am immensely grateful, but also to all the people who have simply posted stories on the internet. I hope my footnotes give the credit where it is due.

As with any book, the following is very much based on my interests and reading. It is not comprehensive and there are aspects of the use of animals, and plants, not covered. For example, I have not included a chapter on the use of sentient species for environmental monitoring, just giving it a short section, as it is a subject well covered by others.

Throughout the book, I use many quotes. These are used so that the exact meaning is not lost, but also because, in some, the phrasing is quite interesting. It should be noted, however, that many appear to have spelling mistakes, or have American spelling. I have endeavoured to keep them original, so such 'mistakes' are deliberate.

Each chapter starts with a fictional story. These are printed in italics to make it obvious that they are not true. I hope that these help to set the scene.

The book starts with an introductory chapter, followed by a chapter on the use of animals to measure air temperature. The next chapter covers the prediction of weather. Earthquakes, volcanic eruptions and tsunami are discussed in Chapter 4, with the fifth chapter looking at the moon, sun and stars. The final chapter attempts to mop up topics not previously covered, such as the use of plants, and will also try to summarise the discussion. Throughout, there is discussion of climate change and how animals may be useful for the future as our environment alters. Much of the detail of the examples discussed is summarised in the tables in the Appendix, which I hope will act as quick source guides.

Lastly, I hope you enjoy this romp through the stories of how animals can be used to measure and even predict our environments.

John T. Hancock
(Professor of Cell Signalling at the University of the
West of England, Bristol (UWE), UK)

Acknowledgements

No book is written in isolation, and this one is no different. I am extremely grateful to all those who have written about this topic, allowing me access to a plethora of stories and information. Some of these were sourced from the scientific literature and others from the internet. Without such materials this book would have been impossible.

As I wrote this book I mentioned it to numerous people, and nearly every time someone had something they wished to add. I had many conversations which continued with: "Do you know about . . .". A good friend, Dr Kirsty Reid, even sent me a message in the middle of the night to tell me that her house in Scotland was being invaded by spiders and that the weather forecast had predicted a storm – she also sent me links to internet articles supplying useful evidence as to why this was significant. To all such friends, I am very grateful.

I would also like to thank the University of the West of England, Bristol, (UWE) for all the resources made available to enable me to compile this book, and for the time given to me to write it.

I am grateful for those who helped and encouraged me to continue writing this book along the way. In particular, I would like to thank my colleagues at UWE, Bristol. Dr David Veal read drafts and suggested alterations, including papers to read and discussed the title of this book. Prof. Jim Longhurst and Dr Mark Everard were also very supportive. Library staff at UWE have been very supportive, and I would like to especially thank Jenni Crossley and Simon Cox. I would also like to thank my brother, Peter, for comments and advising on obtaining images for the book. Unless they are my own, images have been sourced from several places, including Picryl, Wikimedia Commons, the Wellcome Collection, Shutterstock and Unsplash: who I thank for making them available.

I must say a massive thanks to Alice Oven, the editor at CRC Press, who believed in my project and gave me the opportunity to turn it into a book. Thanks to all her team, and especially the anonymous reviewers who also gave supporting advice after receiving a draft of the work.

A massive thanks to all those mentioned and anyone else not mentioned who has persuaded me that this book was a good idea and helped me get to the end.

Lastly, my writing is always supported by my family. My wife, Sally-Ann, and my children, Thomas and Annabel, are so encouraging and help me to the end of large projects.

The author

John T. Hancock is Professor of Cell Signalling at the University of the West of England, Bristol (UWE), UK. In 1984 he was awarded a degree in biochemistry at the University of Bristol, where he stayed to complete his PhD in 1987. Following post-doctoral positions, he moved to UWE in 1993. John has had a long-standing interest in reduction/oxidation (redox) reactions and the molecules involved, but particularly how these mechanisms control cellular function. He has authored several editions of a textbook, *Cell Signalling*, where the processes of how cells perceive and respond to their environment is discussed. John also has several editorial positions for international journals and is Editor-in-Chief of the journal *Oxygen*. Recently, John's research has focused on the role of hydrogen gas in biological systems, and he has written several articles on COVID-19, including about the impact of the pandemic on animals and animal welfare.

CHAPTER

1

Introduction

1.1 Introductory story

The woman was keen to leave. The sky was clear and the wind light. She had promised that she would take her kids to her sister's house for dinner, but it was a long walk up the side of the mountain. She knew that her son and daughter would both object. The path was steep and at this time of year it would be muddy and slippery. However, a promise was a promise, and she knew that her sister would have made an effort to ensure that there was plenty of food. No doubt she would have got treats for the kids too.

"Come on," she shouted up the narrow stairs. "We'll be late." She opened the front door and then abruptly stopped. A small flock of gulls was whirling overhead, but they were inexorably heading inland. The woman turned to look out to sea. A few fishing boats circled around below her, probably being instructed by the huer on the cliffs, looking down to see where the best fishing spots were. The horizon was clear and sharp. There were no signs of any clouds looming from the southwest, the direction of the prevailing weather.

"Put on your stout boots and a raincoat," she instructed the children.

"But it's lovely and sunny," her son replied, grumpily.

"Storm on the way," the woman replied, but gave no explanation of her assertion.

"But-"

"Just behave and come," she said sternly. Her son pulled on his heavy jacket and stomped towards the back gate.

It took them nearly ninety minutes to reach her sister's house. All of them were hot and sweaty by the time they arrived.

"You look as though you could do with a cool drink," the sister greeted them. "Why the rain gear?"

"Storm on the way," the woman repeated.

"You doing your own forecasting again?"

"Gulls. They never lie."

DOI: 10.1201/9781003343844-1

"Right!" the sister said, laughing.

As predicted the dinner was amazing considering that the whole family was poor. A young lamb had been slaughtered, and vegetables straight from the hillside garden were crisp and fresh. But as the kettle started to boil there was a loud crack of thunder. It was going to be a wet and cold walk home.

1.2 Predicting our environment

It is not uncommon for people around the world to be watching the weather. Weather can dominate our lives. What do we wear? Is it going to rain? Do I need sun cream? Can I plant my crops? Do we need to batten down against the coming storm? Will the boats need to be moved as the sea is about to get rough and anything left on the beach will be washed away? The questions could go on, and will be different depending on your location or which people you are with. But wherever you are in the world, the weather can have a profound effect on human activity.

It is not only the weather, but other environmental factors which can have a significant influence on people's lives. The damage caused by an earthquake, or the onslaught of a tsunami, can have effects which last for years, or decades. The eruption of Vesuvius wiped out Pompeii and the surrounding region, now a world-renowned archaeological site. The population and communities there never recovered. The future of similar regions in the world may be secured only if such catastrophic events can be predicted and mitigated.

It is not strange to hear people say things about the weather such as, "Change is on the way," or even "Close tonight." The latter would be a prediction of a storm arriving, usually imminently. It appears that there is something which we are sensing that predicts the change in weather, but such phrases are often muttered in a sarcastic manner or with a sense of cynicism.

Humans are animals, but we seem in general to have lost the ability to use our senses to predict environmental changes, or we simply think our senses are of no significance. Yet we often turn to other animals to predict the changes in the weather. Again, this is often couched in cynicism. When the cows lie down, people comment and laugh. When birds circle, insects run or gnats fly low, people chuckle and take no notice. The chapters that follow attempt to bring such observations together and ask "Is there anything that should be taken seriously here?" Why is it that cultures around the world observe their environment, including the activities of animals, to predict their futures? What are they observing, and, importantly, does it work? Are predictions accurate, or of no consequence? If animals are changing their activities for the oncoming changes in their environment, what are they sensing? Is there a biological, underlying molecular (biochemical) mechanism, which can underpin the changes in animal activity that we see? Why are animals doing it? Is it to their advantage? Can we use animal activity more to predict the weather and other environmental changes? And if we can understand the underlying biochemistry that allow animals to sense the environment, can we build that into the technology of the future?

1.3 Technological and biochemical approaches to measuring the environment

Why, I hear you ask, do we need animals to observe our environment when we have developed, and continue to develop, clever technology-based tools to do it for us? This is a good question. As discussed later, humans have invented an array of such tools. Instruments for measuring the weather include thermometers, barometers and rain gauges. Seismometers measure earthquakes and we have ever-increasingly expensive telescopes for staring into space. Satellites stare down at us, while massive super-computers crunch all the data and give us weather forecasts. There seems little place left here for observing a few animals.

The technologies used in weather and environmental monitoring will not be dealt with in detail here, but some of the history and use of these tools will be covered in consequent chapters. Having said that, if our expectation is that animals can take the place of this technology, they must be able to do the same sensing (or equivalent). If such organism sensing systems exist, they must have evolved from our distance ancestors, and there must be a defined mechanism in these organisms. In some cases, such mechanisms have yet to be defined.

As a working scientist I always think there are two main reasons people like me do what we do. Many will say that it is because of the impact that our work brings to society, and this was certainly brought to the forefront of peoples' minds when many scientists (and others) were applauded for their work on bringing us COVID-19 vaccines with surprising rapidity. However, many scientists get drawn into their world by sheer inquisitiveness. There is a great desire to simply understand how the world works. They will ask questions about the behaviours of animals. Why do certain species have particular activities, and what is the benefit? And if we can understand why animals behave as they do, what can we glean from those observations which is useful to us?

As a biochemist it would be good to know what animals – and plants – can sense. We tend to anthropomorphise our ideas of what other species can do, but there are many aspects of an organism's sensing that we simply do not understand. Much work has been carried out on humans and human tissues, and that is fully understandable as we race to find new drugs and cures for disease. But what do we know about other species? A dip into the genetic databases[1] around the world would make you think there is a huge amount known. Genes for a wide range of organisms are listed. For some organisms, whole genomes – that is all the genes in that organism – are known. But a deeper look will reveal how many of the genes code for 'hypothetical proteins', for which no definite function has been found. Other unknowns include 'orphan receptors', which appear to be sensing proteins, but no one actually knows what they sense. Furthermore, looking for specific genes will quickly show you the limit to the number of species represented. In a recent paper on susceptibility of animals to the SARS-CoV-2 virus (COVID-19), several genes could not be included as they were not in the database.[2] It has been estimated that on Earth there are nearly 9 million eukaryotic species (and there are millions of prokaryotes too), with a startling 86% of land species yet to be described (as of 2011[3]), let alone have their genomes sequenced.

So how do organisms sense things around them? A better question is: how do cells sense what is around them? All cells need to know what their environment is like. Do they have enough nutrients? Are they in the presence of toxins? Are other cells telling them to do a particular activity? In humans, like other organisms, our cells sense their environment with proteins called receptors. These are often embedded in the membrane which makes up the surface of the cell: the plasma membrane (sometimes called the cell membrane). The receptors face outwards and often have several sections, or domains. One of these domains will sense, often bind to, the thing which they are there to sense. There is then a domain which runs across the membrane to the inside surface, where a third domain does something to tell the cell that the receptor has been triggered. Imagine you were at the door of an office. Someone came along and gave you a message for the person in the office. You might read (or listen to) the message (you are acting like an outside domain). You then turn into the office[4] – acting like the domain running through the membrane. You are in the doorway after all, which is effectively part of the office wall (or cell membrane). You will then tell the person in the office what the message was, acting like a domain on the inner side of the membrane. Hence, the message has travelled from the outside of the office (or cell) to the inside, where it can be acted upon. Much of the sensing of cells will rely on such a system. The details may be different – there are several classes, or families, of receptors – but the principles are the same.

Receptor systems are vitally important for the perception of a host of molecules which bring instructions to act to cells. These include hormones, such as insulin, so vital in glucose levels and diabetes when it all goes wrong. But there are a wide range of hormone-like molecules, including the cytokines,[5] which are instrumental in the control of our immune responses, as well as chemokines and pheromones (molecules transmitted between individuals of the same species).

Some receptors are not on the cell surface, but they reside inside cells. A classic example of this is the steroid receptors. Topical steroids are often used for skin conditions, for example, but steroids often get bad press in sports. Many molecules do not use what would be referred to as a receptor system. A good example is the gas nitric oxide, whilst many toxins would be deemed to work in the same manner. However, in all cases, there is a direct interaction with the thing which is being sensed and some component in the cell. This even applies to light, which interacts with proteins in the eye, or the proteins in the leaf of a plant. In all cases, there has to be something which senses the presence of the thing which needs to be detected, and there must be a downstream[6] mechanism to initiate an appropriate response. Without such mechanisms cells would not maintain the correct functions and would not thrive.[7] Without detection and the correct responses, humans would not be able to determine food which is rotten, detect smoke from fires, smell the lion which is about to eat you. Automatically, cells are responding to pathogens and toxins, and we do not even realise it most of the time. All these perceptions and actions are keeping us alive. When it goes wrong, we are in trouble, and a myriad of diseases may result, including cancer. It is worth noting that 34% of drugs approved by the US Food and Drugs Administration (FDA) in 2017 were targeted to one class of receptors, the so-called G-protein coupled receptors. This highlights how important such proteins, and the perception capabilities they bring to the cells, are for the health of the organism.[8]

It is the presence of a receptor-type mechanism which is key to any organism being able to sense what is happening around itself. Therefore, if a bird takes flight, what was it sensing? More importantly, a question which needs to be asked is: does the animal have the capability (receptor mechanism) to sense that thing, be it a sound, light, vibration etc? If the animal cannot sense the environmental cue, it will not respond. It may see other animals responding and decide to do the same, but it would not have directly been sensing the original trigger. In much of the discussion that follows, the exact mechanisms of how animals are sensing the environment is the missing detail. This may be because the mechanism simply does not exist, and therefore the animal cannot sense what we think it can. Or we have yet to discover what is happening. After all, biochemists still have a huge amount to unravel. Just because we do not yet understand does not mean it does not exist.

A couple of examples may be worth a passing note here. There is a growing interest in the use of hydrogen gas (H_2) in both medicine and agriculture. It is said to improve the condition of patients in a multitude of diseases, and it is reported to improve plant growth and crops. And yet how cells perceive hydrogen is not well understood.[9] After all, it is a relatively insoluble and inert gas. Eventually this puzzle will no doubt be solved. In the world of plants, mechanical perception is gaining ground.[10] This is important for how plants grow and manage a mechanical environment, such as the soil. Prince Charles[11] (of the UK) was famously criticised for saying that it was good to talk to his plants,[12] but it appears that plants can perceive sound and respond to it, including music.[13] A biological response to music is often referred to as the Mozart Effect,[14] but exactly what the mechanisms are is not well understood. There has been cynicism to animals predicting the weather and a similar perception of people playing music to their plants. Both may simply be because we are yet to unravel exactly what is happening in these biological systems.

For many, observing animals and trying to decide if it is going to rain is a trivial pursuit. However, for many places in the world, using animals is embedded into peoples' cultures, and has a pragmatic use. Furthermore, if the molecular basis of animal behaviours can be determined, can that knowledge then be used to inform the generation of new technologies?

1.4   Why animals may alter their activities with environmental change

If we can observe the behaviour of animals and then make predictions about our environment, can we understand why are the animals behaving this way? The biochemistry of any organism only has one role. All organisms need to survive, get to reproductive age and then ensure the survival of their offspring. This was nicely summarised by Richard Dawkins when he wrote *The Selfish Gene*.[15] Here, he argued that the organism had no consequence, *per se*, as long as the genes survived and were passed on to the next generation. It was the ongoing persistence of the gene that mattered, not the organism, which was simply a conduit for the gene. The genes, in the form of DNA, are packed in the cells of the organism, regardless of whether they are single-celled, multi-celled, prokaryotic or eukaryotic, as the principles are the same for all. The DNA can be reproduced by the cells' internal machinery, being produced by an enzyme called a polymerase.[16] In eukaryotes this takes place in the cell nucleus, but in prokaryotes it is simply in the cell. Therefore, copies of the DNA, and hence genes, can be given over to the next generation.[17]

Regardless of whether we consider the organism or the gene as the thing that needs to survive and reproduce there are many strategies to do this. Many plants, and even many animals, simply play a numbers game. If a plant produces enough seeds the fact that much of it lands on the tarmac of a road and does not survive, does not matter. Getting crushed or dried out, it makes little difference to the survival of the species as enough lands on fertile ground and survives, so creating the next generation (and the perpetuation of the genes). However, many animals do not have this luxury. Producing too many elephants on the off chance that some may survive is not a good strategy. It is too biochemically expensive. The production of a large individual such as an elephant, which has a low probability of surviving, would simply use too many resources. Producing a very small number of elephants and making sure that, as much as possible, they survive, is a better strategy. Biochemically it makes sense.

In both scenarios the environment can have a major impact. If there are a large number of offspring, spreading them across a range of environments can be helpful, and we often see seeds blowing in the wind. Some of these will land on tarmac, some on the fertile soil. Some may land with no rain coming, whilst other may fall on the edge of a pond, and so can germinate and survive. However, if you only have one or two offspring, and large amounts of resources have been invested in these individuals, they need to survive to reproductive age. Therefore, parents and their offspring will do whatever it takes to thrive, and if that means sensing and acting on changes in their environment then this would be a good thing to do. It is therefore not a great surprise that animals may be risk adverse, behaving perhaps oddly when they sense changes coming. Having said that, not all animals are risk adverse. Some animals are extremely inquisitive, to the point of putting themselves in situations where they might die. Ackerman describes such activity with birds in her book.[18] House sparrows are very good at investigating new environments, particularly those created by humans. A great blue heron (*Ardea herodias*) was described as experimenting with its diet, eating an elasmobranch (in this case an Atlantic stingray (*Dasyatis sabina*)), rather successfully.[19] However, a brown pelican which had tried to do a similar same thing had died,[20] showing that being adventurous is not always an advantage.

Most animals live in the wild, and many are in captivity as pets or in zoos. In the wild, animals are in complex ecosystems and they need to survive. Most human animals – sadly not all – are living in relative luxury compared to many other animals. We know where we can sleep, where breakfast will be, and how to keep dry. If it rains, we put on a coat, or if we get too hot, we take if off. Animals cannot do this, at least in the same timescale – of course many animals moult or grow thicker insulating layers, depending on the climate and time of year. Humans even have ways to mitigate disease, with an ever-expanding array of medicines and vaccines. Many animals have no idea where or when their next meal will be. All too often now we see pictures on the television of polar bears on ever-shrinking ice flows, not knowing where to go and what to do. Are the seals still there and, if not, where is dinner? Such bears may go weeks without food. The imperative in all such situations is to survive and reproduce.

In this hostile environment animals need to survive from birth to reproductive age, and many animals will need to stay around until their offspring are fit and strong enough to survive on their own. Humans are very bad in this regard as organisms, because our offspring are reliant for a very long time. On the open plains an antelope may need to stand and be able to run within minutes of birth. Humans take months, years, to feed themselves. Most animals do not have this luxury. For some leaving the nest is a dangerous time. When a bird first fledges, it often has to be able to fly and escape being eaten if stumbling around on the ground. This similarly applies to bats. Living in a wild environment is a dangerous escapade.

As part of survival strategies animals are likely to become risk averse. If a small bird sees a shadow and thinks that a raptor is hovering overhead, darting away to safety is a sensible thing to do. If that raptor turns out to be an aeroplane, then little is lost, but the bird survives. Do nothing and it may have become something else's dinner, and future reproduction fails to be an option. Therefore, it is not unlikely that animals are constantly observing and sensing their environment. If they sense danger, they are likely to react. Observation of such activity may be useful, but often it may be a waste of time, and such uncertainty plays into the scepticism of using animal behaviour as a useful tool for weather forecasting or predicting an oncoming tsunami.

Many animals are social organisms, living in communities. One survival strategy is to make sure that the whole group knows of the danger and remains safe. Many herd animals try to use their sheer numbers as an intimidation against attack. Others call out to their companions. Birds may call out to warn of danger, or even launch an assault. Recently I have seen both "black birds" (probably jackdaws or crows) and gulls attacking a raptor on the wing above my garden, trying desperately to move it away, I assume from the site of a nest and hence their young – we have a lot of jackdaws in the nearby trees. Elephants will call out in times of distress and to signal danger. There are many other examples across the animal kingdom. Changes in the environment may be the cause of such distress, with unusual animal behaviours ensuing. If an animal is signalling danger to another member of their social group, there is no reason why animals of another species may not pick up on these cues and sense that danger is near. In a garden where there are many species of birds, once one is spooked, they all take to the air. Clearly the sense of danger is picked up by the whole group, not just one species. If other animals can use the observations of other species for their own survival, there is no reason why humans cannot take advantage of this ecosystem-based alarm system too. By observing animals for predicting changes in the weather, or an impending earthquake, is that not what we are doing?

It is likely that many of the observations discussed later are part of such a survival strategy. If a storm is coming, it is best to do what can be done to reduce the danger, or damage to an animal's home, be that burrow or nest. What we might be seeing is a low-risk strategy. Animals act and if danger does not come they have lost little but they have survived. When humans observe this activity and then nothing significant seems to follow, the cynicism seems grounded. However, when a storm hits after the gulls flew over, others will say, "I told you so!" Therefore, many of the observations and anecdotes are hard to take seriously. Here, the following discussion will collect together a range of such observations. In the final chapter, a conclusion will be discussed and, if nothing else, the next time you walk through a cloud of gnats you may consider whether you need to ask why they are there, and it is not just to annoy you!

1.5 Long term, or the here and now

When observing animals, it is worth considering the timescale for which the changes in activity are relevant. Some animal behaviours are simply because they are triggered by the ambient environmental factors present at the time. They are simply reacting to the here and now. Other observations may suggest that something is arriving imminently, in the next few minutes or hours. Yet others might be an indication of what going to happen tomorrow, or even the next day.

The activities that are simply reacting to the present conditions are easiest to understand. The senses of the animal will be reacting to levels of light, or to temperatures. When we say that an animal is restless because there is going to be a catastrophic event in two days' time, such as an earthquake, it much harder to take seriously, and very hard to understand what the animal is perceiving.

It is worth pointing out that although humans have good senses, they are inferior to those of many animals. We have a range of senses: sight, hearing, smell,[21] taste and touch. This involves specialist cells, in our eyes, ears, nose, mouth/throat and skin. However, such sensing cells have limits to their sensitivity. Our eyes can see what we refer to as visible light, and this will include light with wavelengths from approximately 400 nm to 700 nm. This allows us to see anything which is 'visible' – a very anthropomorphic word, as what is visible to us is not the same for other organisms. In our eyes we have two type of specialist cells: rods and cones. The internal membrane of such cells contains rhodopsin, which senses the light. The protein contains what is referred to as a prosthetic group, which is a non-protein part. In this case it is a molecule of 11-*cis*-retinal which is converted to the all-*trans* version by light. This leads to a series of biochemical events[22] which result in the stimulation of the optic nerve, and hence a signal back to the optic centre of the brain. In the brain, the signals arriving are unscrambled to give us a view of the world. Rods are good at picking up low-light levels, but do not do colour. Cone cells, with a slight alteration of the biochemistry, are able to sense colour, but they are not so good at low light. This is discussed further later, when discussing the Northern Lights, which we do not see with the bright colours so commonly seen in photographs. The human eye, because of how it is constructed at a biochemical level, has limits to the light it can perceive. Anything outside of these limits simply 'does not exist' as far as our eyes are concerned and we do not see them. However, animals may be different. Birds, for example, can sense near-ultraviolet light (wavelength 320–400 nm),[23] whilst on the other hand, snakes may be able to sense infrared light, which they use for hunting at night, as they can then see warm-blooded animals in the dark.[24]

The range of sound frequencies which are sensed by humans is also limited. It is commonly said that humans have a range of approximately 29 to 20,000 Hz, although individuals do vary and the high-pitched sound perception drops off with age. In a similar manner to light perception, hearing does not have the same limit in animals. Elephants for example, can sense sound at lower frequencies than humans, down to 17 Hz or even below, and they can hear 'calls' from around 2.5 km away.[25] On the other hand, bats are well known for their sound perception which they use as radar. Whilst watching bats drink on the wing from a swimming pool in Catalonia I completely trusted the bats' radar as they skimmed past my face in the dusk. I was later told, by an academic who is a bat expert, that I was lucky not to be hit, as often even bats can get it wrong. Bats appear to use a range of sonic frequencies in their hunting strategies, and this alters as they approach their prey. Frequencies between 5 Hz and 200 Hz may be used,

A big brown bat, hanging from a tree.
(Sourced from Shutterstock, contributed by Ovidiu Dugulan. Image has been made black and white and cropped.)

as reported for the big brown bat, (*Eptesicus fuscus*). I certainly did not hear any sound emanating from the bats that whizzed past my ears, but they were clearly using something as they skimmed the water to drink and then dodged me standing in their flightpath.

There is a danger, therefore, to anthropomorphise our thoughts when we are wondering how animals are perceiving factors in the environment which are not available to us. There are clearly a range of things in the environment which we do not see, hear or feel. It is these senses which we need to embrace if we are to understand the behaviour of animals, and also need to consider if we are ever to embed such senses into our future technologies. It is also such senses which may allow animals to predict events which appear to be well in advance. The behaviours of animals may start to alter well in advance to humans sensing anything, either using their own senses or technology available to us. Later the perception of earthquakes will be discussed, and some say that prediction of such events is impossible, and will remain impossible, at least for the foreseeable future. So, what is happening in seismometer zoos in China, used to predict earthquakes well in advance of any perceivable rumblings of the Earth's crust?

Therefore, throughout the discussion that follows, the idea of what the timescale of behaviour change will be attempted to be clarified. Are animals responding to the here and now, or are they truly predicting the future? Certainly, as we shall see, thousands of lives have been saved in a variety of situations by observing animals, allowing humans to mitigate disaster which would have caused the loss of life. The animals were simply protecting themselves, but the upside of observing them is of possible benefit to humans too. And obviously the longer the warning an animal can give means that we can act in good time before disaster strikes. It takes time to abandon a city, but less time to leave a house or turn the gas-main off. If animals are truly giving a long-range warning, then perhaps less cynicism is needed and we can learn from how animals react to what we cannot see coming.

1.6   Piquing my interest

The practice of watching animals to determine the weather and other environmental events, such as earthquakes, has been carried out for centuries. Some of the anecdotes are hard to comprehend, and many are the subject of mirth. Are there any truths to any of these myths and sayings, and are they underpinned by science? Over the past three decades of reading and discussing aspects of biochemistry, I have become intrigued by this rather underrated area of science. If I told people in the 21st Century that the gulls are flying inland and therefore rain is on the way, they would no doubt pick up their smartphones and tell me that I am talking rubbish. However, not everyone has access to such technology, and it is frequently wrong, or simply unable to predict what is about to happen. My smartphone has two weather apps and they are often telling different forecasts. We are perhaps too reliant on technology, and we are losing some of the old skills that our ancestors used. Maybe it is time to revisit some of these stories and determine if there is any reason to take notice of them.

My interest probably started when I stumbled across Dolbear's Law, but events in one's life also have a profound effect. In fact, events are not always remembered for the reasons which would have been predicted before they took place. The event itself may have been fantastic, but something else lingers in the memory, and it might not be obviously related to the star turn.

The date was 11th August 1999. I was heading to Cornwall with my family for a truly once-in-a-lifetime event. It was going to happen at 11:04:09 in the morning, exactly. Miss it and it would not be repeated in the UK until 23rd September 2090.[26]

Cornwall (or Kernow, in Cornish) is an interesting place. It has its own language (Kernewek) and many longstanding cultural events, such as the Helston Flora Dance and Padstow's Hobby Horse. It has its own cuisine, exemplified by the Cornish pasty and saffron buns,[27] but accompanied by other local foods such as Cornish Yarg (a cheese[28]) and even an ongoing dispute about whether jam or cream goes on a scone first (jam first, the Cornish way, is correct, despite what they say in Devon![29]). The coastal regions are beautiful, with rugged cliffs and long sandy beaches, and attract tens of thousands of tourists every year. Surfing is popular with wide waves often rolling in, especially on the north coast. The inland regions are scoured by the numerous mining industries which included the extraction of tin, copper and even arsenic, particularly in the west of the county. The old granite engine houses are famous around the world. The central area of the county has been turned into opencast mines and piles of white sand from the clay industry. I can still remember when the rivers ran bright white from the outfall of these mines, with some of the beaches being formed by the sand that was washed down to the sea, such as at Crinnis.[30] Such mining industries led to Cornish engineers and scientists becoming world renowned, such as *Richard Trevithick* (1771–1833)[31] who invented the first high pressure steam boiler and then the first steam engine to run on a track (called Catch Me Who Can[32]), amongst other things, and Sir *Humphry Davy* (1778–1829),[33] who invented the mining safety lamp, again, amongst many other things. His prowess also included writing poetry. However, none of these features of the county were forefront in the minds of the thousands of people who were flocking over the Tamar.[34]

We joined the queues on the A30 and headed to my parents' house. I was born in St Austell (mid-Cornwall), and went to school there, only leaving to go to university.[35] But well before the important moment approached, we drove up the hill to get a better view of the main event. We parked off the by-pass and stood on a grass bank, giving us a commanding view of St Austell Bay. We could see the Gribbin Head promontory to the left and had a view out towards Trenarren in the west. In the front of us was the urban sprawl of Holmbush and Carlyon Bay behind that. The weather was a little touch and go, with some cloud, but we were hopeful. At least we had the best chance, having such a wide view out over the English Channel.

As the time approached, we could see the darkening of the sky approaching from the west. Gradually the sun was obscured, and we were plunged into the gloom. It was not totally dark, but it was much nearer to night-time than daytime. It was quite eerie. The atmosphere was quiet and noticeably cooler. But the inexorable happened, and the dark shadow of the moon moved east as the elements of the solar system continued on their

Diamond ring effect of a solar eclipse.
(Original photograph by Matt Nelson on Unsplash. This picture has been changed to black and white.)

never-ending travels. The 'diamond ring' was created, the light brightened and then it was all over, or at least the main event.

We left our viewing spot and returned to my parents' house where we continued to watch as the moon tracked across the sun and eventually continued on its way, not to be so popular until the next eclipse. Cornwall too had had a moment of fame, with pictures broadcast around the world. The news was replete with stories and photographs that day.

But what has this to do with the topic of this book? In front of our vantage point on the grass bank was a field. Agriculture in Cornwall is mainly based on animals, with little level ground for swathes of wheat or barley. In the field were horses, which is not uncommon in the St Austell area. They had no knowledge of the impending event. We were not close and were in no way disturbing them. But all of a sudden, as we were patiently waiting for the eclipse, the horses started to run up and down the field. They looked spooked. They were clearly not happy. Something had upset them. At this time there was no sign of the eclipse; we had arrived very early in case lots of other people decided that our vantage point was a good one, and we were patiently waiting ahead of time. There was no obvious sign of anyone else around to disturb the animals. As far as I could tell, the temperature had not dropped and there was no darkening of the sky. Even so, the horses had sensed something amiss. Therefore, what was upsetting them? Did they somehow sense that the eclipse was coming? Were they more sensitive to changes in the environment than I was? Whatever they were sensing was not to their liking, obviously. And lastly, if they were sensing a forthcoming eclipse, could other animals also sense such events, and if so, can us humans use such knowledge as predictors of environmental changes?

It is one of those events which sticks in the memory. I have thought often about those horses, as much as the eclipse itself. The eclipse was indeed a once-in-a-lifetime event, and I was glad we made the effort to be there. But the behaviour of those horses, whether caused by the arriving eclipse, or just a coincidence, also reminds me of the other animal behaviours which people used to point out when I was younger. "The gnats are flying low tonight, rain is coming," was often said, particularly by my Mum, who as far as I could tell had a long Cornish ancestry.[36] Rain is very common in Cornwall, as the peninsula sticks out into the Atlantic and the south westerly prevailing (often wet) winds. I never did work out if this saying about the gnats was true, but this idea will be revisited in Chapter 3. As I was growing up, we were, as already mentioned, not far from the south coast, and if gulls were flying inland, we were told that this was a warning that a storm was coming. Perhaps the gulls had been out to sea and seen the weather, but I am sure that the gnats had not flown ahead of the rain. Therefore, if such animals are indeed predictors of the changing weather, what is it that they are sensing that I am not? I am an animal too, after all. Furthermore, is there a logical reason why they need to sense the altering environmental conditions? As discussed above, does it help in their survival and therefore their reproductive potential? After all, all species need to survive only long enough for the next generation to be created. Beyond that, growing old for most species is simply a burden. However, if the next generation is not produced, the species would die out. A biochemistry professor at my old university used to joke that he had done his job now as he had grown-up children and therefore from a biological point-of-view he was worthless. Despite his candour, monitoring the environment and changing activity or adapting to that environment is crucially important to species' survival. A quote often attributed to Charles Darwin is: "It is not the strongest of the species that survives, nor the most intelligent that survives. It is the one that is most adaptable to change."[37] As mentioned already, Dawkins argued that successful reproduction allows the survival of the genes,[38] and this drives the need for all organisms to thrive and reproduce, and evolve so that they don't get out-competed and die out. All such observations and thoughts started to whirl around my brain, and when I read about Harlow Shapley in Sobel's book entitled The Glass Universe,[39] I decided it was time I did something about gathering all these ideas into one place. My interest was truly piqued, and hence, here we are.

1.7   Relevance and climate change

I am, of course, not the only person to note the relationship between animals and changes in the weather. In fact, I am not the first to bring some of these myths and anecdotes together in one place. In 1991 a book was published which was supposed to be written by "Uncle Offa".[40] Uncle Offa was from a radio show, the BBC's *Farming Week*. He was first broadcast at 6.46 am on the last day of September 1990, on BBC Radio 4. Uncle Offa was, in fact, Frederick Hingston, but he took the nom-de-plume as he was from a village "within a Sabbath Day's journey of Offa's Dyke". Offa's Dyke is an earthwork which follows the border between Mercia and Wales, and was named after an Anglo-Saxon king (ruled AD 757–796).[41] Hingston attempted to collate any thoughts and sayings which may help forecast the weather. Hingston served in the Devonshire Regiment in many countries around the world, but later became a freelance writer. The book was illustrated with cartoons by Leonard C. Gilley. Gilley was a cartographer in the Royal Air Force during World War II, and went on to work for several large companies, including Shell and Cadburys, as well as working with several publishing companies, including MacMillan.

It appears that the radio show was contacted by numerous people with sayings and myths, and these are the basis of the book Hingston put together. In Uncle Offa's book, which is written in a rather flippant and jovial manner, there is mainly references to particular days which can be used for the weather to be predicted. For example, "Look at the weathercock on St Thomas's Day [21st December], the wind will remain there for three months." However, some of his sayings are a little odd – although no fault of his as he was just collecting sayings which were known. He mentions how Easter can be used to predict the weather, but of course Easter is not a fixed date, rather it follows the cycle of the moon.[42] But of relevance here, Uncle Offa also mentions animals in several places. In the Introduction of his book, he points out that in China weather forecasting often involves children: from a quote from the *Daily Telegraph*, 19th October 1987, Offa quotes:

> The pupils, whose predictions are reputed to be extremely accurate, obverse the local weather patterns and any unusual behiviour by animals.

On page 133, Uncle Offa even has a short section on earthquake predictions. This, and Offa's examples of animals and weather predictions, will be discussed in the relevant chapters later.

As will be seen, the correlating the observation of animals with the environment has been written about for hundreds of years, and is embedded in numerous cultures. There is a myriad of myths, sayings and anecdotes. For example, in *The Old Farmer's Almanac* there are several phrases given:[43]

> See how high the hornet's nest, 'twill tell how high the snow will rest.
> If ant hills are high in July, the coming winter will be hard.
> When cicadas are heard, dry weather will follow, and frost will come in six weeks.
> If ants their walls do frequently build, rain will from the clouds be spilled.
> When bees to distance wing their flight, days are warm and skies are bright; But when their flight ends near their home, stormy weather is sure to come.
> The early arrival of crickets on the hearth means an early winter.
> The more quickly crickets chirp, the warmer the temperature.

Such biological reports are intriguing. Clearly animals can sense their environment and act accordingly. Of pertinence here, over the course of history, the reverse has been suggested too. The observing of animal behaviour can be used to measure and even predict the environment in which we live. But do any of these myths and anecdotes have any relevance to the 21st Century? After all, we have modern technology and can measure a wide range of environmental conditions, including what the weather is doing at any point in the world and predict what it might be like in several days' time. My smartphone gives its GPS coordinates and then claims to give me a bespoke weather forecast for wherever it is located, and it will also give me simultaneous forecasts for anywhere else too – useful for travelling so I know to wear lots of layers if heading north of the Arctic Circle, for example. Why do I need animals to do this for me? Perhaps I don't, but there are many places on the globe, and many environmental events, for which animals can be useful. And if nothing else, it is intriguing science which can tell us much about the biology of the animals around us.

As I continued to write this text, the 26th UN Climate Change Conference of the Parties (COP26) had started in Glasgow, Scotland, UK[44] (31st October–12th November 2021). Such an international event, at which all the major world leaders were expected to attend,[45] showed that this was a truly important moment. Countries were expected to confirm, or announce, their pledges to make their societies carbon neutral, limiting the emissions of carbon dioxide (CO_2) and therefore limiting the warming of the global surface temperatures. It was hoped that the ambitions of the Paris Agreement,[46] (decided at COP21 in Paris on 12th December 2015, by 196 parties), where the global temperature rise should be kept below 2 °C, compared to pre-industrial temperatures, and preferably a rise of below 1.5 °C, would be met. Many countries have outlined their intentions to be "zero carbon" by limiting their use of fossil fuels such as natural gas, coal and oil, and offsetting use by mitigating schemes. Alternatively, countries are putting in place green alternatives, such as wind farms (multiple wind turbines), solar panels or nuclear power stations, moving their power needs over to sustainable electricity production. For example, the UK government committed to become carbon zero by 2050.[47] Other countries have been much more reticent, with many hoping that Australia will commit to a better climate change policy.[48] By December 2021, politicians in Europe were arguing whether gas (a fossil fuel) should be included in a taxonomy of sustainable fuel sources.[49] This is said by some, such as Greta Thunberg,[50] to undermine the agreements only very recently agreed at COP26. Indeed, this would cast into doubt the ability for the global temperature rise to be restricted to +1.5 °C, or even +2 °C.

With global temperature rises estimated in some modelling to now reach higher than 2 °C unless human behaviour changes[51] there are bad future outcomes forecast. These include loss of ice in the polar regions,[52] melting glaciers[53] and extreme weather events.[54] All these discussions emphasise how important it is to observe and understand our environments. Understanding the weather, and how it may change due to anthropogenic-influenced climate change, is vitally important now and will only increase in significance in the future.

With the 2022 Winter Olympics taking place while I wrote,[55] a recent report really highlighted how climate change may have an impact, and also be impacted upon. It has been estimated that out of the last 21 locations around the world that have been used for the Winter Olympics, only one of those will be suitable as a reliable place in the future, if global temperatures rise as predicted.[56] Others give similar gloomy estimates. For example, the BBC says:

> Of the 20 Winter Olympic venues since 1924, scientists believe only 10 will have the 'climate suitability' and natural snowfall levels needed to host an event by 2050.

Such news was hammered home when it was pointed out that the 2022 Olympics in Beijing was the first ever to use nearly all artificial snow.[57,58] Although fake snow has been used extensively before, including in many ski resorts (it has been suggested that 95% of resorts use snow making[59]), the use of water and energy in Beijing is an environmental concern. Not only was water diverted from other uses, such as needed by the city (it is estimated that 222 million litres of water are needed), but the energy use no doubt contributed to climate change. Having said that, ironically, the Men's Giant Slalom at Beijing was disrupted and was made very difficult due to heavy natural snow fall! It has, however, been argued that such sporting events need to be given better consideration in the future, so that the environmental impact is less. Climate change will exacerbate such issues, but these events will continue to contribute to the problem. Other sports, such as football and motor racing, may need to change the way that they are run and supported, as has already been suggested.[60,61]

To emphasis the impact on the world's climate, newspapers in March 2022 reported that the polar regions were much warmer that they should be, both in the north and south. In the Arctic, the temperature was 30 °C above average for the time of year, whilst the Antarctic was 40 °C above what it would normally be.[62] The Arctic is leaving its winter, whilst in the south winter is on the way. Both extremes seem to be affected.[63],[64] This will have significant effects on ice melting, increases in sea levels, and hence more flooding, and changes in weather patterns around the globe.

As discussed in the final chapter, both animals and plants will need to adapt to these inevitable changes in global temperature and the extreme weather events that will ensue. The latter part of 2021 has seen extreme tornadoes in the USA, for example. These were described by some as a "rare event called a derecho",[65] or an inland hurricane. Such events, described as rare, are becoming more common. In 2021 alone, record-breaking snow falls were brought to Madrid by Storm Filomena, whilst Storm Christoph brought heavy rain and flooding to North Wales in the UK, and it was described as "one of the wettest three-day periods on record".[66] The same website lists 13 other extreme weather events in 2021, including floods in India and Nepal, as well as in China and Germany; wildfires in Greece (caused by dry weather and heat); and a heat dome in the northwest of the USA, which was described as an event that "rewrote the record books".[67] In July 2021, the *Guardian* newspaper ran a podcast on such extreme events and why they are becoming more common.[68] Such news reports emphasise three important points. Firstly, such events are becoming more common and more extreme. Secondly, this is a global issue, with reports from numerous countries in disparate places on Earth, from the USA to Japan. No place seems immune. Even the Arctic regions are not safe, with temperatures causing major issues, including for the animals which live there.[69] Thirdly, the notion of climate change is getting more traction in the press and becoming more embedded in the consciousness of many people, not just those who study it. Whether this knowledge and awareness manifests into tangible action which limits the global temperature changes is yet to be seen. As 2021 closed, the *Guardian* reported the hottest New Year's Eve on record being followed by the hottest New Year's Day on record for the UK.[70] The records of extreme of weather seem to continue to tumble, all around the world, and this trend is very likely to continue.

Of pertinence here is the fact that is it is not just humans that are victims of these events. Animals may be able to sense such forthcoming events and respond appropriately. This might mean finding deeper water in which to shelter, running to a forest, or simply flying off to a safer place. However, before the animal gets hit by a storm, it has to sense that it is coming. It is no good waiting and then struggling to survive the impact. By that time, it may be too late. It is much better to predict impending doom and do something about it. Therefore, can we ascertain if animals are able to sense environmental changes to which they can respond in order to survive? If we can, can we then use that knowledge to our advantage and predict future catastrophic events? Can we extend this knowledge to predict not just weather, but major events such as earthquakes?

As we shall see, animals do appear to respond to the environment around them and have been used for predicting imminent rain, or dry periods, in numerous cultures for generations. Such knowledge is going to be ever more important to understand. Some of the myths and anecdotes are bizarre and hard to ground in any scientific context, but others make sense and do have a rationale which can be used to understand why animals behave as they do. Because some of these observations and sayings have endured the test of time, it is evident that even if we don't understand why, some of the animal-based predictors of weather and climate are interesting to consider. As the climate changes, it will also be interesting to know how environmental-inspired animal behaviours change and how such alterations of animal activity can help us understand how we are changing the world around us.

1.8   Chapter summary

Using the observation of animals to predict the weather has been around for hundreds of years. Born in 371 BC, Theophrastus wrote about examples where this could be useful. Using animals in this way may not be new, but there is much scepticism about this topic. Many people just giggle if someone points to a herd of sitting cows. However, clearly animals must constantly observe their environments and act accordingly. The imperative for all organisms is to reproduce and pass on their genes, enabling their species to continue into the future. Being cognisant of danger is a major part of the survival strategy of many organisms. Mitigating the arrival of a storm may be the difference between life and death.

The environment is changing. A major concern in the early part of the 21st Century is climate change. Although the Earth's climate has constantly changed through geological time, rapid changes now being recorded are exacerbated by human activity, such as burning fossil fuels and releasing 'greenhouse gasses' into the atmosphere. The governments of many countries are vowing to put in place measures to slow the rising global temperatures, but some governments refuse of partake, or circumstances do not allow as rapid a change in policy as many would like to see.[71]

Numerous technologies are in place, and will be further developed, which can measure many of the aspects of the environment that are discussed here. However, it can still be argued that observing the antics of animals has a place in giving a holistic view of what may be happening, either now or in the near future. As climates change, ever more rapidly, observing other organisms may be very valuable to determine what the future may bring.

My interest in this subject was piqued by personal observations and many examples I found in reading a range of books, from fiction to biochemistry. Clearly, animals do alter their behaviour according to their environments, and even changes to the environment that they perceive. What is also clear is that a wide range of famous and brilliant people also took the time to contemplate this topic and write about it, including Theophrastus, Sir Humphry Davy, Edward Jenner and Harlow Shapley.

Of course, it would be naïve to suggest that just using animals would solve all our problems in predicting the environment of the future. In the final chapter there will be a brief discussion of plants, but of course many cultures use a wide range of predictors. Modern technology, developed over centuries of invention and experimentation, offers hugely important tools, and they will be briefly discussed in relevant chapters. However, even some of these immensely expensive instruments can be outdone by an animal when predicting the environment. It seems that it is timely to evaluate the evidence.

Endnotes

1 One of the major sources of genetic data is the National Center for Biotechnology Information (NCBI). Here the sequences of genes and proteins can be found, and there are many tools available for deeper analysis. www.ncbi.nlm.nih.gov/ (Accessed 18/06/22)
2 Hancock, J.T., Rouse, R.C., Stone, E. and Greenhough, A. (2021) Interacting proteins, polymorphisms and the susceptibility of animals to SARS-CoV-2. *Animals*, 11, 797.

3   Mora, C., Tittensor, D.P., Adl, S., Simpson, A.G. and Worm, B. (2011) How many species are there on Earth and in the ocean? *PLoS Biology*, 9, e1001127.

4   Action of receptors relies on what is referred to as a conformational change. The protein will change its shape in the perception of the thing it is sensing.

5   Cytokine signalling can go very wrong. Many have died during the COVID-19 pandemic due to what is referred to as a cytokine storm, where dysfunction of the cytokine system overcomes the normal control mechanisms of cells.

6   Biochemists often refer to molecules or regions of molecules (in DNA and RNA) being upstream (before) or downstream (after). This is particular prevalent when discussing genes and signalling pathways.

7   It is worth noting that some signals trigger the proactive death of the cell, in a process called apoptosis (cell suicide). This would have happened as you developed, for example, to prevent you having webbed fingers.

8   For a more in-depth discussion of cell signalling see: Hancock, J.T. (2021) *Cell Signalling*. Oxford University Press, Oxford. ISBN: 9780198859581

9   Hancock, J.T. and Russell, G. (2021) Downstream signalling from molecular hydrogen. *Plants*, 10, 367.

10 Toyota, M. and Gilroy, S. (2013) Gravitropism and mechanical signaling in plants. *American Journal of Botany*, 100, 111–125.

11 As I edit this manuscript, he has become King Charles III.

12 Telegraph: www.telegraph.co.uk/gardening/5080991/Prince-of-Wales-talked-to-plants-scientists-test-if-it-works.html (Accessed 30/08/22)

13 Chowdhury, A.R. and Gupta, A. (2015) Effect of music on plants–an overview. *International Journal of Integrative Sciences, Innovation and Technology*, 4, 30–34.

14 Jenkins, J.S. (2001) The Mozart effect. *Journal of the Royal Society of Medicine*, 94, 170–172.

15 Dawkins, R. (2016) *The Extended Selfish Gene*. Oxford University Press, Oxford. ISBN: 9780198788782. This is the longer version; the original was published in 1976 and launched Dawkins' career. It has remained a seminal text ever since.

16 It is the internal machinery of cells that viruses "hijack" and so make more copies of their DNA/RNA and new "offspring". As an example, having been through the COVID-19 pandemic, such biochemical mechanisms have global relevance and need to be understood.

17 Copying and passing on DNA is not infallible, and mistakes do often happen. Mutations (changes in the genetic code) of an animal (or plant or microbe) can occur, perhaps as a simple polymerase mistake or forced, for example by radiation exposure. Chromosomes undergo what is known as recombination, and genes can be transferred 'horizontally' in many species. Without genetic changes in offspring evolution could not happen. Offspring are not simple identical copies of a parent, else we would all look exactly like either our mum or dad.

18 Ackerman, J. (2016) *The Genius of Birds*. Corsair, London. ISBN: 9781472114365 [Chapter 8]

19 Ajemian, M.J., Dolan, D., Graham, W.M. and Powers, S.P. (2011) First evidence of elasmobranch predation by a waterbird: Stingray attack and consumption by the great blue heron (*Ardea herodias*). *Waterbirds*, 34, 117–120.

20 Bostic, D.L. and Banks, R.C (1966) A record of stingray predation by the brown pelican. *The Condor*, 68, 515–516. https://doi.org/10.2307/1365329

21 It is worth noting that work on human smell was awarded the Nobel Prize in 2004. The work was carried out by Richard Axel (Columbia University) and Linda Buck

(Fred Hutchinson Cancer Research Center, Seattle). The BBC: http://news.bbc.co.uk/1/hi/health/3713134.stm (Accessed 16/06/22)

22 Hancock, J.T. (2021) *Cell Signalling.* Oxford University Press, Oxford. ISBN: 9780198859581

23 Rajchard, J. (2009) Ultraviolet (UV) light perception by birds: A review. *Veterinární medicína,* 54, 351–359.

24 Newman, E.A. and Hartline, P.H. (1982) The infrared "vision" of snakes. *Scientific American,* 246, 116–127.

25 Jacobson, S.L. and Plotnik, J.M. (2020) The importance of sensory perception in an elephant's cognitive world. *Comparative Cognition & Behavior Reviews,* 15.

26 Royal Museums Greenwish: www.rmg.co.uk/stories/topics/eclipses (Accessed 22/06/22)

27 Rather unusually, saffron is used a lot in Cornwall in buns and cakes. Cornwall Guide: www.cornwalls.co.uk/food/saffron_buns.htm (Accessed 22/06/22)

28 Yarg is a hard cheese coated with the leaves of the stinging nettle. It was said to be invented by a person called Gray, hence the name. Lynher Dairies Limited: https://lynherdairies.co.uk/lynher-dairies-home/yarg-got-name/ (Accessed 22/06/22)

29 The Spruce Eats: www.thespruceeats.com/difference-cornish-vs-devon-cream-tea-435316 (Accessed 22/06/22)

30 Crinnis is a long beach at Carlyon Bay, in the centre of St. Austell Bay. Cornwall Beaches: www.cornwall-beaches.co.uk/austell-riviera/carlyon-bay.htm (Accessed 22/06/22). The beach itself is mainly sand and grit from the washings of the local clay industry, unlike most of the others which are natural sand.

31 Burton, A. (2000) *Richard Trevithick: Giant of Steam.* Aurum Press Ltd., London. ISBN: 1854107283

32 Catch Me Who Can: www.catchmewhocan.org.uk/home.html (Accessed 22/06/22)

33 Lamont-Brown, R. (2004) *Humphry Davy: Life Beyond the Lamp.* Sutton Publishing, Stroud, Gloucestershire. ISBN: 9780750932318

34 The Tamar is the river which separates Cornwall from the rest of England. It nearly, but not quite, dissects across the peninsula.

35 Cornwall had no university until Exeter University opened Cornish campuses. Of course, Camborne School of Mines had been around since 1888.

36 As did my Dad.

37 Quote Investigator: https://quoteinvestigator.com/2014/05/04/adapt/ (Accessed 09/08/22)

38 Dawkins, R. (2016) *The Extended Selfish Gene.* Oxford University Press, Oxford. ISBN: 9780198788782

39 Sobel, D. (2017) *The Glass Universe: The Hidden History of the Women Who Took the Measure of the Stars.* 4th Estate, London.

40 "Uncle Offa" (1991) *Natural Weather Wisdom.* The Self Publishing Association Ltd., Worcs, UK.

41 Offa's Dyke can now be followed using a path, which is approximately 177 miles (285 Km) long. It roughly follows the English/Welsh border. National Trails: www.nationaltrail.co.uk/en_GB/trails/offas-dyke-path/ (Accessed 29/08/22)

42 Easter is determined to be the first Sunday after a full moon on or after 21st March. Time and Date: www.timeanddate.com/calendar/determining-easter-date.html (Accessed 29/08/22)

43 Almanac: www.almanac.com/how-insects-predict-weather (Accessed 22/06/22)

44 UN Climate Change Conference UK 2021: https://ukcop26.org/ (Accessed 22/06/22)

45 President Putin of Russia and the President Bolsonaro of Brazil did not attend, and President Xi Jinping of China promised to attend via a digital platform, although also did not attend. This was significant as they are representatives of high fossil fuel using countries.

46 United Nations Climate Change: https://unfccc.int/process-and-meetings/the-paris-agreement/the-paris-agreement (Accessed 22/06/22)

47 UK Parliament: https://lordslibrary.parliament.uk/climate-change-targets-the-road-to-net-zero/ (Accessed 22/06/22)

48 GlobalCitizen:www.globalcitizen.org/en/content/australia-climate-crisis-cop26/?gclid=EAIaIQobChMIp8GZ6L3y8wIVEeztCh2NiQAlEAAYAiAAEgKBLvD_BwE (Accessed 22/06/22)

49 The Guardian: www.theguardian.com/world/2021/dec/21/eu-in-row-over-inclusion-of-gas-and-nuclear-in-sustainability-guidance (Accessed 22/06/22)

50 Greta Thunberg is a young climate activist, who came to note whilst she was still a teenager. She has been very outspoken about protecting the planet from climate change. The BBC: www.bbc.co.uk/news/world-europe-49918719 (Accessed 21/12/21)

51 Global Climate Projections: www.ipcc.ch/site/assets/uploads/2018/02/ar4-wg1-chapter10-1.pdf (Accessed 22/06/22)

52 Global Citizen: www.globalcitizen.org/en/content/global-ice-loss-worst-case-scenario/?gclid=EAIaIQobChMIqbPqgcDy8wIVgmDmCh0gTwSxEAAYAyAAEgK7UfD_BwE (Accessed 02/11/21)

53 Global Citizen: www.globalcitizen.org/en/content/himalayan-glacier-disaster-india-climate-change/?gclid=EAIaIQobChMIrcn5r8Dy8wIViKztCh36vQsvEAAYAyAAEgLPL_D_BwE (Accessed 02/11/21)

54 United Response: www.unitedresponse.org.uk/resource/easy-news-extreme-weather-climate-change/?gclid=EAIaIQobChMIlevC4MDy8wIVB7TtCh0nTgolEAAYASAAEgJmHfD_BwE (Accessed 22/06/22)

55 4th February to 20th February 2022. Beijing 2022: https://olympics.com/en/beijing-2022/ (Accessed 22/06/22)

56 The Guardian: www.theguardian.com/environment/ng-interactive/2022/jan/25/rising-temperatures-threaten-future-winter-olympics-games-global-emissions (Accessed 25/01/22)

57 The BBC: www.bbc.co.uk/sport/winter-sports/60130999?at_medium=RSS&at_campaign=KARANGA (Accessed 29/01/22)

58 IFLScience: www.iflscience.com/environment/this-years-winter-olympics-will-be-first-ever-to-take-place-on-100-percent-artificial-snow/ (Accessed 07/02/22)

59 The Newstatesman: www.newstatesman.com/environment/2022/01/beijings-green-winter-olympics-looks-as-fake-as-its-snow (Accessed 29/01/22)

60 Game of the People: https://gameofthepeople.com/2022/01/12/non-league-football-can-lead-the-way-in-the-climate-agenda-2/ (Accessed 22/06/22)

61 The Guardian: www.theguardian.com/sport/2021/nov/26/climate-emergency-accelerates-f1-efforts-to-clean-up-image (Accessed 29/01/22)

62 The Guardian: www.theguardian.com/environment/2022/mar/20/heatwaves-at-both-of-earth-poles-alarm-climate-scientists (Accessed 21/03/22)

63 The Guardian: www.theguardian.com/environment/2022/mar/21/extremes-of-40c-above-normal-whats-causing-extraordinary-heating-in-polar-regions (Access 21/03/22)

64 Euractiv: www.euractiv.com/section/climate-environment/news/heatwaves-at-both-of-earths-poles-alarm-climate-scientists/ (Accessed 21/03/22)

65 The Guardian: www.theguardian.com/us-news/2021/dec/20/us-tornadoes-storms-derecho-national-weather-service (Accessed 21/12/21)

66 The Week: www.theweek.co.uk/news/environment/953574/worlds-most-extreme-weather-events-2021 (Accessed 22/06/22)

67 The Week: www.theweek.co.uk/news/environment/953574/worlds-most-extreme-weather-events-2021 (Accessed 21/12/21)

68 The Guardian: www.theguardian.com/science/audio/2021/jul/20/why-are-extreme-weather-events-on-the-rise-part-one-podcast (Accessed 21/12/21)

69 Earthwatch: https://earthwatch.org/stories/trees-tundra?gclid=Cj0KCQiAk4aO
BhCTARIsAFWFP9Gx2FBle6ZHqstdw22-5OZtIUlGWgHZGApJGeOhid5S
dgM0kfEbEioaAqPeEALw_wcB (Accessed 22/06/22)

70 The Guardian: www.theguardian.com/uk-news/2022/jan/01/warmest-uk-new-years-day-follows-record-breaking-new-years-eve (Accessed 03/01/22)

71 As I write this a war in Ukraine is having a profound effect on supplies of fossil fuels, as well as food, across the region and beyond. Some climate change policies are being watered down or reversed, to allow life to continue as normal as possible in many countries.

2

Measuring ambient temperature

2.1 Introductory story

"That was a cracking day!" Brian announced.

"A hard walk, but worth it," Alice agreed.

"The views were amazing, right down the gorge. Amazing!"

"Certainly worth the trek."

"I'll let the fire die back, and then we can turn in. Judging from the weather today, it's going to be a hot night."

"Actually, I think we need to keep it going," Alice replied.

"Why? We've finished all the cooking, unless you need even more coffee?"

"A nice glass of wine is what I need! But the air's cooling. It could be a cold night."

"What do you mean? I'm sweating here."

"Listen!" Alice instructed.

"What am I listening to? All I can hear are those confounded crickets."

"That's the point. How many chirps are there per minute?"

"What? Have you been on the wine already?"

"They're slowing down. If you can't feel the temperature, then they can."

"Here, have more wine," Brian offered Alice a glass of red.

"After I've collected some more wood," Alice ignored Brian's outstretched hand.

Ten minutes later, Alice had brought over a pile of logs. "There, that should keep us warmer."

"Here, drink the wine, before the crickets do."

"Dolbear's Law, Brian. You'll see."

"I'll drink to your eccentricity, but it's going to be a hot night."

"You'll see. I believe the crickets. They know. I'm going to get an extra blanket, do you want one?"

DOI: 10.1201/9781003343844-2

"Not for me, thanks," Brian smiled, shaking his head. However, by the middle of the night he was wishing that he had not been so rash. In the dark and cold, he fumbled around looking for a blanket that he could use too.

2.2 Measuring the air temperature with an animal – an introduction

In 1897 Amos Dolbear published a paper entitled *The Cricket as a Thermometer*, where he posited that listening to the chirps of a cricket can be used to calculate the ambient air temperature. Despite this becoming known as Dolbear's Law, he was not the first to suggest or even publish the idea. Regardless, such work does suggest that there is a direct interaction between the behaviour of an animal and its environment.

If animals can be used to measure, or predict changes to, the environment they need to be able to perceive that environment and have an alteration in behaviour that can be seen and understood by us. One of the many atmospheric variables which we use to monitor, report and forecast the weather is the air temperature, and here Dolbear suggested the use of animals. So does the ambient temperature alter the way animals behave and, if so, is Dolbear right and can this be used as a measure of the temperature? In other words, can we use this as a proxy thermometer? Furthermore, does this make any sense and can the mechanism underlying such observations be determined?

There are many intriguing animal behaviours associated with environmental conditions, and many are hard to understand. Whilst at a conference in Germany many years ago, I was told that leaf-cutter ants (such as *Atta sexdens* or *Acromyrmex lobicornis*) respond to the levels of oxygen available. These ants are truly social animals who work as teams. Ants cut sections off the leaves and then carry them back to the underground nest. There the leaf material is used as a fertiliser to grow a fungus (*Leucoagaricus gongylophorus*), which supplies nutrients on which the ants feed. These ants are acting as farmers. It has been suggested that a full colony of ants can use as much plant material as a cow.[1] The academic telling us about this said that the ants inside the nest are smaller than those working outside. This was because there is less oxygen inside the nest, so their respiration had evolved to be appropriate. Insects receive their oxygen down tubes known as tracheae, with the ones deeper in the tissues being smaller and referred to as tracheoles. The oxygen available to the insects' cells is therefore dependent on the movement of gas down these tubes. Dawkins suggested that the whole of an insect's body can be considered to be a lung.[2] The longer the tubes, the less oxygen can get to the tissues. If oxygen is low on the outside, this lowers the potential amount of oxygen reaching the cells. Therefore, insects in low oxygen would be smaller, reducing the gas exchange distance. This makes sense and is exemplified by the reverse, seen during the Carboniferous period. The atmospheric oxygen was then around 35%, as opposed to 21% today.[3] This allowed the evolution of insects which were much bigger than we see today, and this was epitomised by the giant dragonfly. Such animals could survive because of the higher oxygen, which could diffuse down the extra-long tracheae and allow the cells to still have enough oxygen to drive their aerobic (oxygen-dependent) respiration. Observing that the cutter ants inside the nest were smaller, where the oxygen concentration is less than 21%, therefore makes sense. However, the academic went on to say that the size of the section of leaf that was cut by the ants was also dependent on the oxygen concentration. This also makes sense. Less oxygen means less efficient respiration and less energy, and therefore cutting a smaller leaf section which needs to be carried back to the nest would be a sensible thing to do. Physiologically it is logical. But how does the ant know what to cut before starting? If what this academic said is true, the ants clearly can sense the oxygen tension and make an informed decision. They never seem to get halfway back to the nest, decide the leaf is too heavy and cut a bit off. They cut the leaf and go, so have to be able to deliver the load.[4] Theoretically, if this relationship between the size (weight?) of the leaf section cut by the ants was plotted against the oxygen concentration at the time, then any future assessment of the size of leaf cut by the ants could then be used to measure the oxygen available. The ants would, in fact, be acting as an oxygen meter. To my knowledge, this has never been suggested, but if the academic in Germany was correct, then such relationships between the behaviour of an animal and environmental conditions, such as oxygen, can be used to our advantage – albeit that electronic oxygen sensors also exist. As we shall see, this may never have been suggested for measuring oxygen, but it *has* been suggested, and used, to measure other environmental conditions. Leaf cutter ants also have been suggested to predict future weather,[5] but this will be revisited in Chapter 3. However, in this chapter too, ants will be prominent.

2.3   Importance of measuring temperature

It is clear that animals can sense their environments and respond accordingly. To be able to sense the weather, and even predict the weather which may arrive, animals need to sense a range of environmental conditions. One of these is the air temperature. However, what is quite extraordinary is that the response of the animal can be used to measure the air temperature. The animal can be used as a thermometer. One useful animal here is the ant.

Temperature and its measurement are immensely important to today's society. Several times a day, news broadcasts tell us what the air temperature is and what it is likely to be, either in the imminent future or over the next few days, or even weeks. Similarly, newspapers give forecasts. On a relatively trivial note, we wish to know whether to go to the beach or to go skiing. On a more serious note, as the world's climate changes, there is more urgency about knowing how hot the atmosphere is, and how we may stop it getting hotter.[6] Icecaps are melting, glaciers sliding off mountains,[7] and there are regions of the world where air temperatures are eye-watering. More days are being recorded as being over 50 °C.[8] This is a hard temperature to imagine, let alone suffer. Death Valley in the USA recently recorded 54 °C (130 °F).[9]

In a medical arena, the measurement of temperature is also hugely important. One of the first things that a medic will do is record your temperature, and in hospital this is regularly monitored. A rise in temperature can be an easy-to-measure sign that something is wrong. One of the results of an inflammatory response is a rise in body temperature.[10]

Measuring temperatures around the ambient air levels is therefore important. No one would expect any biological material to be useful below freezing or above the temperature which is not survivable – proteins denature as the temperature rises, after all, as exemplified every time you fry an egg. However, measuring air temperatures is useful and here animals may be able to help. On the other hand, several instruments have been developed to do this job, and they are easier to use, more sustainable and more accurate.

2.4 Instrumentation which has been developed to measure temperature

To measure either air temperature, or that of a human body, one would turn to a thermometer, not to an ant. It is said that the thermometer was invented by Galileo Galilei,[11] as depicted on an Italian postage stamp printed in 1945. He created what was known as a thermoscope. In 1612 a scale was added to a tube of expanding liquid by Santorio Santorio, who was a Venetian scholar. In the 1650s, Ferdinando II de' Medici, of Tuscany, used a sealed tube and created something more akin to the thermometer as we know it. It was independent of air pressure, so was a significant improvement. A better scale was proposed in 1701 by Olaus Rømer, a Danish astronomer. He used a scale from brine[12] to boiling water. The

Italian postage stamp depicting Galileo Galilei (1564–1642), 1945.
(Sourced from Picryl. This image has been made black and white.)

Marble sculpture of Galileo by Pio Fedi. It was installed in the Lanyon Building in 2001 at Queen's University Belfast, Belfast, Northern Ireland.
(Photo by K. Mitch Hodge on UnSplash. This image has been made black and white.)

polish scientist Daniel Gabriel Fahrenheit, in collaboration with Rømer, invented the mercury-based thermometer and also created the Fahrenheit scale, still commonly used today. Fahrenheit (F) would be generally used in the USA today, for example. However, in 1742, Anders Celsius, a Swedish astronomer, created the scale where boiling water was at 100 degrees, but the French physicist Jean-Pierre Christin altered the scale so it ran from 0 to 100 (frozen water to boiling). Weather forecasts usually use Fahrenheit or Celsius (otherwise referred to as Centigrade: C), depending where in the world you are. Here in the UK Celsius has been adopted, although I remember Fahrenheit being used when I was younger. Celsius is most commonly used across Europe. Despite F and C being the usual scales people are most accustomed to and would be expected to be used on a TV weather forecast, from a scientist's point of view, the temperature scale was named after William Thomson. He was originally from Belfast but moved to Glasgow.[13] He became Lord Kelvin, and we now have the Kelvin temperature scale (K), which starts at absolute zero (0 °K: −273.15 °C).[14] Temperatures below this are not possible.[15,16]

Although the liquid-based thermometer is still hugely important, and regularly used in domestic, medical and industrial settings, today there are digital thermometers which are much more sensitive and easier to use. The use of thermocouples was suggested by an Estonian scientist, born in Tallinn, called Thomas Seebeck (1770–1831).[17] He realised that two metals in close contact had magnetic properties which change with temperature.[18] This is known as the Seebeck Effect. Hence, the early digital thermometer was born. In 1964 a German physician called Dr Theodore Benzinger designed a thermometer which could take a person's temperature using their ear, and this technology is in common use today. There are, therefore, numerous instruments which can be drawn upon to measure the temperature, either of a solid object or the air.

Yet – who needs a thermometer when we have ants?

2.5   Measuring temperature with an animal

Humans, like many animals, are warm-blooded. Our body temperature is around 98.6°F (37°C). However, it is not exactly fixed and can fluctuate slightly. Normal body temperatures tend to range from 97°F (36.1°C) to 99°F (37.2°C).[19] One of the ways in which temperature is maintained is through the use of brown fat. Cells within this tissue have a high respiratory rate and instead of converting the energy into chemical energy which can be stored or used (for example, to move muscles), the energy is 'lost' as heat.[20] This process is called thermogenesis (the generation of heat) and uses a protein known as thermogenin.[21] In organelles called mitochondria in cells, there is an electrical gradient generated across the inner membrane. This is produced by the movement of protons 'out' of the mitochondria, into the intermembrane space (mitochondria have two membranes the protons do not leave the organelle). Usually, the protons return to the inside (matrix) of the of the mitochondria through a protein complex which generates adenosine triphosphate (ATP) from adenosine diphosphate (ADP) and inorganic phosphate. The ATP generated by this mechanism is used to drive muscle contractions and other metabolic processes, regenerating the ADP which can then be recycled.[22] The protein thermogenin allows the protons to return to the inside of the mitochondria with no ATP production, and the energy which would have been 'entrapped' in the ATP is lost as heat. There is some anecdotal evidence that people who struggle to put on weight, or do not feel the cold, have relatively more brown fat. There is also similar evidence that brown fat can be increased by prolonged cold exposure. There seems to be little robust evidence for either of these.

In the very young, thermogenesis is important. Brown fat lines the vital organs such as the spinal cord, ensuring such tissues are kept warm. This is especially important in the young as their surface area to body mass is relatively low, and their heat loss relatively high. The surface area, where heat loss will occur, of a small body (e.g. baby mammal) is relatively large compared to its cellular volume. In an adult the surface area is much lower compared to the body volume, even though the actual surface area is much larger. Therefore, relative heat loss is lower.[23] On the other hand, physiological responses such as sweating will keep the temperature down, so that body temperature is maintained within a relatively narrow range.

In humans, a temperature over 100.4°F (38°C) most often means you have a fever caused by an infection or illness, hence why it is so closely monitored in hospital. On a molecular level, it was found that when mammalian cells are heated to temperatures which are too high, but not high enough to actually kill them, there is the production of a set of proteins known as the Heat Shock Proteins (HSPs).[24] It was assumed that these were produced as a stress response, which they are, but they are also used in a range of normal physiological activities, including protein folding, protein movements in cells and protein degradation. Hence, warm-blooded animals have clever ways to control their body temperature and respond if such regulation goes awry.

Cold-blooded animals, on the other hand, are dependent on the environment to maintain their body temperatures. Often reptiles can be seen sunning themselves on the top of rocks (in Cornwall, adders[25] can be seen basking on granite slabs in the summer), to increase their body temperature and therefore their metabolism. Reptiles can also get too hot[26] and

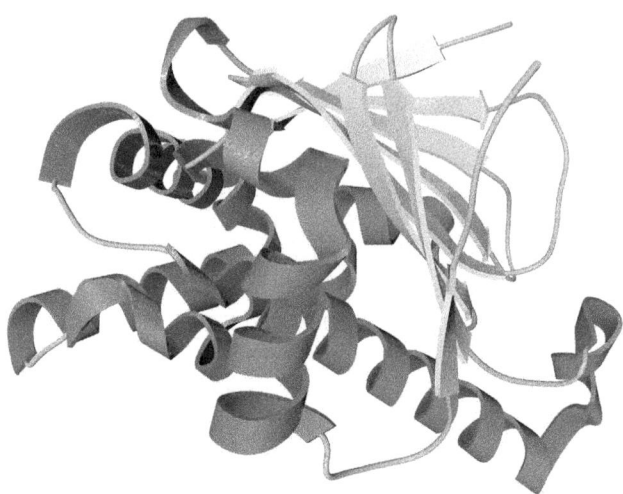

A representation of the structure of a heat shock protein (HSP 90a-NTD bound to adenine: data entry 7s90). This is shown as a ribbon diagram, where arrows are a structure called a beta-sheet, and ringlets are alpha-helices. This was deposited at the Protein Data Bank by T.R. Stachowski and colleagues, where hundreds of such representations of proteins are freely available (www.rcsb.org/).
(Image sourced from Shutterstock, contributed by Maryna Olyak, but made black and white.)

are sometimes seen looking for shade or even diving into water to cool off. Therefore, their activity is very dependent on ambient temperatures. Cold-blooded animals work hard to maintain their body temperatures, even if it looks as though all they are doing is lazing on a rock. They are also, it should be remembered, quite vulnerable whilst they are basking in the open, and hence in the sunshine, so these animals are often very vigilant. If you try to approach an adder on a Cornish rock, the snake will not defend its position but will rapidly disappear before you can get too close, even if it appears that it was not watching you approach.

As a rule-of-thumb, the rate of a chemical reaction will double for every ten degrees (°C) rise in temperature.[27] Of course, this will be altered by the catalytic rate of enzymes, and any temperature above protein denaturation (usually mid 50s °C) will see the enzymatic rate plummet. However, it is unlikely that a cold-blooded animal could get that hot (although temperatures, as discussed earlier, can get into the mid-50s (°C) in some places in the world). Reptiles have heat shock proteins too,[28] so can respond to high temperatures at a molecular level. Even so, as the temperature of cells in a cold-blooded animal rises so do the metabolic rates, including the respiratory rate. Therefore, cool cold-blooded animals are sluggish, making them easy prey or poor predators. Warm cold-blooded animals are spritelier, being able to escape or capture, and therefore much more likely to survive. Hence, cold-blooded animals sun themselves and work hard to maintain an appropriate temperature, neither too cold nor too hot, rather keeping their cells in a Goldilocks zone.

Therefore, as insects are cold-blooded too, as an ant gets warmer, it can run faster!

Harlow Shapley (1885–1972) was a world-famous astronomer and became the head of the Harvard College Observatory (1921–1952).[29] He is probably most well known for his suggestion that the sun was not the centre of the Milky Way, as previously thought. He originally studied journalism at the University of Missouri, but then moved to do his graduate studies at Princeton University, where he worked under Henry Norris Russell. He visited Harvard in 1914, where the then director of the institute, Edward Charles Pickering (1846–1919), offered Shapley help. Subsequently, at Mount Wilson Solar Observatory, Shapely used the 60-inch reflecting telescope, and his work on the Capheid variables, which are stars which pulsate, enabled him to realise that the Milky Way was bigger than previously

Progressive Citizens of America (1947) party members. Left to right, seated, Henry A. Wallace and Elliott Roosevelt; standing, Dr Harlow Shapley and Jo Davidson.
Photo from the New York World-Telegram and Sun collection at the Library of Congress.
(Sourced from Picryl.)

suggested, and that the sun was not in the centre. He was brilliant at his work, and his name is amongst the greats of astronomy.

However, like many brilliant people, Shapley's mind never seemed to stop enquiring. It was at Mount Wilson that the activity of ants caught his eye. He started to notice how their behaviour altered as they went under a manzanita bush. As they went into the shade, they ran more slowly. This clearly intrigued him. Having assumed that, like us, the ants would appreciate some respite from the heat, he decided to do some recording. In fact, this side-line obviously became an obsession, as he wrote in his memoir: "Observing [the stars] was always hard for me." He put this down to his extramural activities. "I 'suffered' quite a bit those long cold nights. I suppose I didn't get as much sleep in the daytime as I needed for I was running around observing ants in the bushes."[30] However, such daytime activity was not wasted, but became a scientific endeavour in its own right. Shapley now had two scientific projects. He studied the stars during the night and ants during the day.

Shapley's first scientific paper[31] on the behaviour of ants appeared in 1920.[32] This paper starts off precisely summing up what was happening:

> Variation in the activity of a cold-blooded animal is largely dependent on metabolic changes, which in turn probably depend mainly on the accelleration of oxidation and of other chemical reactions. The physical nature of the environment affects most of these chemical processes, and we should expect that the same physical properties would also directly influence the kinetic manifestations of animal life.

This paragraph shows that he was in fact spot on. All he needed now was the experimental evidence. And being a great scientist, that is what he set out to achieve.

It was clear that he was not just observing one species of insect either. He says, "the most suitable material to be the Argentine ant, *Iridomyrmex humilis* Mayr, and two species of *Liometopum*". He had also investigated using *Tapinoma* and *Dorymyrmex* (both types of ants), but found them not so conducive to study. It is also interesting that he obviously researched insect behaviour, as he comments that trail-running is quite common in *Dolichoderinae*, but not for *Leptomyrmex* of Australia. Such behaviour allowed Shapley to set up his experiments knowing that the ants would be running the same route. He pointed out that the same trails had been used by ants at Mount Wilson for two years. He also pointed out that this use of the same trail by ants was not uncommon, having been reported in Europe. He suggested that the ants used the trail probably for patrols as they were not carrying anything (unlike the leaf-cutter ants that were discussed earlier), which would of course have altered their speed. Therefore, he had an ideal subject for his studies.

The site that Shapley used allowed him to observe the animals over a wide range of environmental conditions, including temperature, humidity, wind and light. The temperatures ranged from 38° C (over 100° F) down to "less than 8°". Relative humidity ranged from 100% down to 5%. He even reported that thousands of ants could be observed "only a few feet from banks of snow". He could also observe the ants during the day and at night, when changes in light could be taken into account. One has to assume that his night-time observations took him away from his telescopic studies on those days when the ants attracted his attention more than the stars. Clouds, of course, are an astronomer's worse enemy, so he may have had plenty of time if the stars could not be seen.

Shapley also noted that the ants were "carnivorous, granivorous, and aphidicolous" and he could use this to his advantage if he wanted to maintain their running, or to start new trails. However, little encouragement seemed to be required on warm days, as shown by the sheer scale of his observations. He reported that on a good day 100 ants per minutes would pass a given point, and rather amazingly:

> From one nest on Mount Wilson, in the summer of 1919, four files issued, and every day, under favorable conditions of weather, 70,000 ants went out along each file and nearly 70,000 ants came in.

Nine separate speed traps were set up, but two of them became his preferred choice. One was at the base of a concrete wall and the other on a plank of wood. Whilst the plank was near the nest, the wall was about 50 feet away, but Shapley pointed out that both were of similar smoothness and length, being about 30 centimetres long. The whole experiment was relatively close to an officially maintained weather station, so as well as knowing how long the ants took to go through his speed trap, Shapley could also record the temperature (near the trail with a thermometer), along with precipitation, wind-velocity and barometric pressure. He also recorded the size of the ants and their direction of travel. It appeared to be a very thorough experiment, and he did not appear to have any preconceptions.

It quickly became apparent that the major influence on the speed of the ants was temperature. The light levels and time of day had little effect. Rain, he pointed out, did have an effect, but this was not unexpected (rain in the film *A Bug's Life*[33] showed how animators considered rain to affect insect life, with huge raindrops dropping like bombs). With repeated observations Shapley tabulated his results, and with so many measurements he could do the appropriate statistical analysis. The data of temperature against speed was created, combining the recordings from his two preferred speed traps. He even included two data points he thought were erroneous, and then explained why they should be ignored (discussed later).

From a large number of observations and recordings, Shapley came up with five concluding points. The first, and most important, is that temperature and the speed of the ants are related, as can be seen from the data he published:

> From the measurement of the speed of a thousand individuals of the species *Liometopum apiculatum*, an empirical curve is obtained that for any temperature throughout a range of 30° centigrade gives the speed with an average probable error of 5% for one observation.

He had established that there was a clear relationship between the activity of the ant and ambient temperature, but then this information could be used to actually measure the temperature. In the same paragraph as the preceding quote, he claimed that "from a single observation of the ant-speed, the temperature can be predicted within 1 degree centigrade". This seems

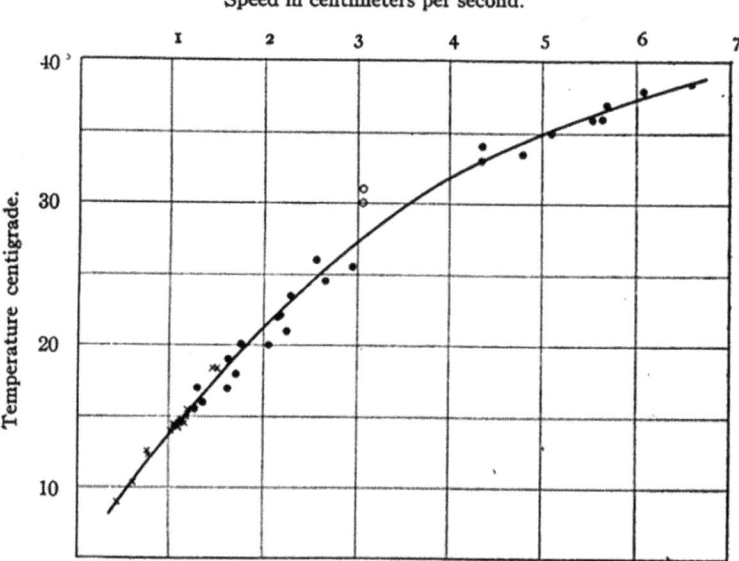

Fig. 1. The temperature-speed curve for *Liometopum apiculatum*.

Data from Shapley's original 1920 research paper: Shapley, H. (1920) Thermokinetics of *Liometopum apiculatum* Mayr. *Proceedings of the National Academy of Sciences of the United States of America*, 6(4), 204.

remarkable. I am sure he didn't mean a single ant here, but a determination of the average speed recorded at that temperature. Even so, to be able to say that the ant speed gives this level of accuracy is a bold statement. Secondly, the data are more rigorous at the higher temperatures. Thirdly, the speed increased 15-fold for a 30° C rise in temperature, "increasing uniformly from 0.44 to 6.60 centimeters a second". This, of course, is no surprise, as his own opening statement in the paper said that there was a relationship between the "physical nature of the environment" and metabolic activity. Fourth, the speed of the ants was more or less the same, regardless of the direction in which they were running, either towards or away from the nest. As the ants were not carrying anything and seemed to be able to run unhindered, this too is not a great surprise. Lastly, the ants were as active in the night as during the day, and between 14° C and 38° C there was no difference in the number of ants which were active.

One of the most amazing observations Shapley made he gave as almost a throwaway comment. Whilst explaining why some data didn't fit the curve, he commented that "the artificial lowering of temperature is responded to much more quickly by the ants than by the mercury thermometer". This was a surprise to me. I would expect that a change in metabolic activity of cells or enzymes in an *in vitro*[34] experiment would behave like this, but to be seen *in vivo*, in a whole animal, does seem quite amazing. The cells involved are cocooned under an exoskeleton and not directly exposed to the air, so some lag may be expected. The warm (or cooler) air would need to track down the trachea and bathe the cells, and then allow the cells to equilibrate at that temperature. Insects are small, but I would still expect a lag, but Shapley didn't seem to see such a significant delay. Perhaps his mercury thermometer was particularly slow to respond?

The observations made were explainable when it is considered that the ants are cold-blooded and Shapley knew this, giving this explanation as the opening of his paper. But what was remarkable was that the data fitted a curve so neatly and then could be used in reverse, that is, to measure the ambient temperature once the speed of the ants is known.

Shapley didn't stop there but went on to publish a follow-up paper in 1924, also in *Proceedings of the National Academy of Sciences of the United States of America*.[35] He was obviously proud of his achievements with ants, as he was quoted in saying, "One of the most interesting points I have gathered in my scientific career is the speed at which one particular kind of ant will run with a rising temperature."[36] Considering his enormous reputation as an astronomer, this is an astonishing statement. In this second paper he reiterated his findings in the first paper, saying, "Observation of the time required to run a distance of thirty centimeters, taking an average for ten or twenty individuals, suffices to indicate the air temperature within one degree." He expanded his work, studying a variety of ant species: "*L. occidentale, L. luctuosum* (or a similar undescribed species), *Dorymyrmex pyramicus* subsp. *brunneus, Iridomyrmex analis, I. humilis,* and *Tapinoma sessile.* All these forms belong to the subfamily Dolichoderine." Shapley reiterated how useful such species were, as they ran trails, were active both night and day and formed large colonies. He found that all the species investigated showed the same temperature dependence, "but only for *T. sessile* and the Argentine ant, *I. humilis,* were the observations sufficiently systematic to merit numerical presentation".

He didn't wish to stop there either. He went on to try to determine why this run speed to temperature relationship existed, calculating Q_{10} (temperature coefficient) values and invoking the works of van't Hoff, Arrhenius, Volkmann and Krogh.[37]

Interestingly, the ant run for *Tapinoma* was through a room, and therefore Shapley could easily control the temperature. However, there was a limit to how warm he could make the room, and the ants stopped entering and became impossible to measure. He said that if the temperature "became higher than the optimum, the ants refused to enter the room. Those already in the room at temperatures higher than 40° became erratic and sluggish, and no longer passed through the speed trap".

The observations were taken in "1920 on August 4, 5, and 6 for *Iridomyrmex* and on September 5 and 6 for *Tapinoma*", but it was four years later before the paper was published. Clearly this was a side-line, fitted in when he had the opportunity, and he appeared not to publish anything further on this subject, returning to his astronomy. He was director of Harvard College Observatory from 1921 until 1957. This may account for his loss of interest in, or time to devote to, ants. He also had five children, all born around this period of ant observations.

Shapley's biological ventures were not ignored. The second paper was quoted 37 times.[38] For example, there was a paper in the *Transactions of the Royal Entomological Society of London* in 1931, by B.P. Uvarov,[39] and Shapley's 1924 paper is still being cited nearly one hundred years after its publication.[40] Shapley's first paper faired a little better, being cited 57 times. The idea of measuring the running speed of ants and how this is altered by temperature is still being studied in the 21st Century, for example by Hurlbert *et al.* in 2008.[41] Behaviour studies of ants in the face of climate change even quote Shapley's work.[42] In 1959 the muscle proteins of ants were investigated[43] and data compared to Shapley's findings. The authors of this paper said:

> As a result a quantitative identification between the physiological process and the purified enzyme are possible over a wide temperature

range. The agreement is within experimental error at every temperature. This provocative correspondence is not obtainable with any other enzyme whose temperature dependence is known.

Clearly, Harlow Shapley's observations of ants has borne the criticism of time and are still being cited over one hundred years since his chance observation and first publication.[44] A doctoral thesis written in 2021[45] states, "Temperature directly influences walking speed (Shapley, 1920." One can see why Shapley was proud of this aspect of his scientific work, and it no doubt will be quoted well into the future. He gave a legacy in astronomy and a second, smaller and less influential, one in biology.

Ants were obviously quite an obsession of Shapley's anyway. In an interview with Helen Wright and Charles Weiner, at his home in New Hampshire on 25th August 1966, Shapley talks about how he collected ants from all over the world.[46] At a visit to the pyramids in Cairo, Egypt, he complains that he had left his vials behind and had no way of pickling his finds. He therefore borrowed Detlev Bronk's watch and put an ant inside:

> We unscrewed the crystal and put one of the ants in underneath the crystal, and shut it up. The ant crawled over to five minutes of eleven and perished; so I had the beginnings of a menagerie with me.

On another occasion he put ants in his tobacco pouch. A little later, during a game of bridge, he got annoyed by one of the other players:

> It annoyed me; wouldn't it annoy you to have a good bridge game spoiled by this foul ambition of a British officer? So I hauled out my jimmy pipe and calmed my nervous anger with a solacing pipe. But I forgot what I had put in that particular tobacco pouch. Later I found that alas, alas, my ants had all been smoked away.

And then he mused, "It was sad because I'd carried the ants so t[f]ar with me." Clearly, ants and their antics were very close to Shapley's heart, and it was not just at Mount Wilson where he studied them. It seems a pity that there appears to be no other papers which he published on the subject.[47]

The relationship between the behaviour of animals and temperature is not confined to ants. Many cold-blooded animals have a similar relationship. This would be expected. There is no biological reason why this phenomenon should be restricted to ants. However, other observations seem to be based on the noise made by the animals, rather than their run (movement[48]) speed.

Rattlesnakes are a group of venomous snakes. They belong to the genera *Crotalus* and *Sistrurus* (subfamily *Crotalinae*). They are all classed as vipers. Their name comes from the rattling sound which emanates from their tails, which is used as a warning and to deter predators. However, it has been observed that the rate of the rattle is dependent on the ambient air temperature. In a paper by J.H. Martin and R.M. Bagby[49] in 1972, the abstract states:

> The rattling frequency of the rattlesnake, *Crotalus atrox*, does not appear to be linear throughout a body temperature range of 3–40° C, although it does appear to be linear between 16 and 32 C.

The authors go on to say that above 34 °C the tail muscles on either side, which work antagonistically, contract at the same time and therefore the movement of the tail becomes spasmodic. Therefore, the characteristic rattle is not heard.

The work was carried out on ten snakes supplied by Otto Locke from Texas, with the authors working in the Department of Zoology and Entomology in Tennessee. They used two methods for measuring the temperatures of the snakes. The first was a thermistor inserted 9 cm into the cloaca (sewer in Latin), an opening into which the intestines and urinary tract deposit. Secondly, the animals were fed a thelemetry probe. Temperatures could therefore be recorded to the nearest 0.1 °C, which is more accurate than Shapley would have managed with a mercury thermometer. The animals were placed into a cold room, at 4 °C,[50] and then the room was slowly (a rise of less than 0.2 °C per minute) warmed to 40 °C with lamps. The frequency of the tail rattle of each animal was recorded for at least 50 temperatures. The authors also recorded the size (weight and length) of the animals, their molt state and the animals' health. To induce the animals to rattle a blunt stick was used, and the rattle was recorded with a variety of means. But it was noted that the rattle frequency was not invariant at any temperature, but the highest frequency was least variable. This seemed to agree with Shapley who noted that his data were most accurate at the higher temperatures and therefore the highest rates of activity. To attempt to get measurements at low temperatures the animals were placed on crushed ice, but rather unsurprisingly the snakes did not seem very happy and could not easily be persuaded to rattle. For animals which would sit in the sun and try to warm themselves, being on ice must have been a bit of a shock.

The paper shows the raw data as a figure, and the central section is amazingly linear. Between 8 °C and 36 °C the authors calculated a regression for the best fit of $R = 155T - 283$. R is the rattle cycle per minute, and the temperature is represented by T.

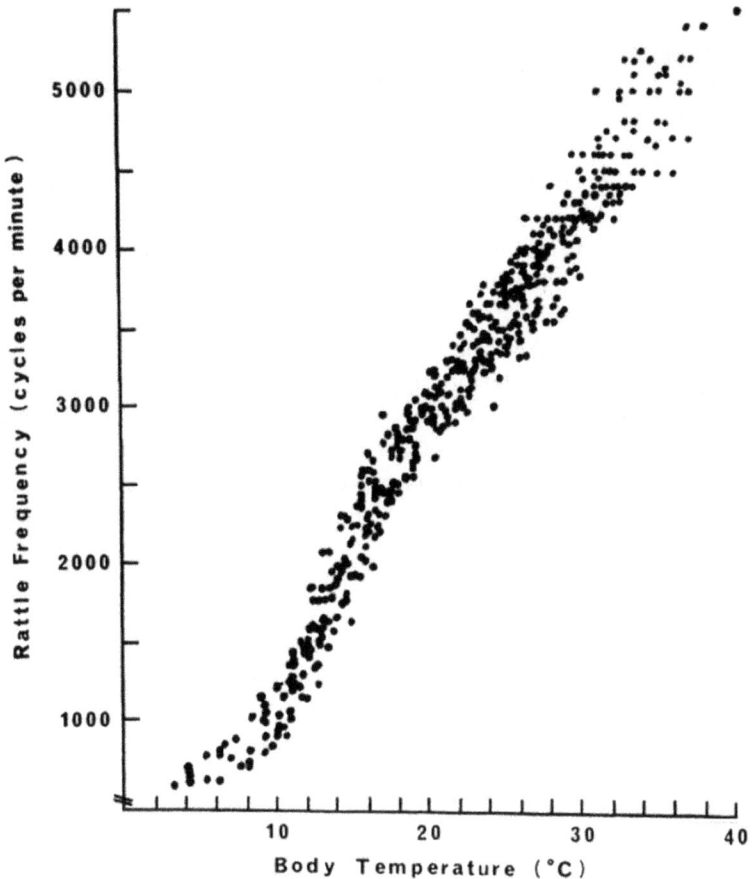

The original data from Martin and Bagby's original paper: Martin, J.H. and Bagby, R.M. (1972) Temperature-frequency relationship of the rattlesnake rattle. *Copeia*, 482–485.

Interestingly, this was not the first observation of the dependence of the rattle frequency of snakes and temperature, having been published in *Science* in 1954.[51] L.E. Chadwick and H. Rahn had used a different species of snake, that is, *Crotalus v. viridis*. However, Martin and Bagby said that their data were consistent with this earlier work. The data by Chadwick and Rahn appeared to be collected in 1941, using 18 snakes. The animals were placed in an incubator at 37 °C or in an ice box at 5 °C, and left there to acclimatise for one to two hours. Rattle frequency was assessed using a mercury arc stroboscope light. Animals were left to warm or cool, as relevant, and the temperature measured with a thermometer inserted at least

6 cm into the cloaca. The data they showed as a figure was a straight line from approximately 2 °C to 40 °C, although both the ends are extrapolations. Both Martin & Bagby and Chadwick & Rahn were not content to just show the correlation data but they furthered the experiments by inserting electrodes into the muscles and looked to see if they could ascertain what was happening. In these papers they could see which muscles were working, and why the tail has the oscillation which creates the rattle. It also showed that at high temperatures the action potentials[52] on either side of the tail overlapped, and hence the movement was sporadic and the lack of a noticeable rattle.

Studies such as these show that the frequency of sound emanating from the snake's tail has a relationship to temperature, therefore this, like the speed of ants, can be used in reverse. If the rattle frequency is measured, then the body temperature of the animal can be determined quite accurately, and as the animal is cold-blooded this can give a good estimate of ambient temperature. Martin and Bagby noted that they used a different rattlesnake species, compared to the earlier study, so checked with other rattle snake species too. Unsurprisingly, they too showed the same relationship. Therefore, this phenomenon can be used for whichever rattlesnake is encountered, assuming the rattle frequency can be accurately measured. At 36 °C this is a rate of over 5,000 cycles per minute, so some instrumentation would be needed. It should be noted, that carrying a thermometer is easier and, dare I say, safer. However, even so, it shows that the relationship of ambient temperature to measurable biological activity is not limited to ants.

However, Chadwick and Rahn were not the first to publish a paper on this topic either. A paper was published in 1940 by L.M. Klauber.[53] Chadwick and Rahn say that Klauber:

> observed in *Crotalus v. viridis* an average increase of 93 cpm/° C between external temperatures of 11 and 22⁰ C. In Klauber's measurements, the motion was recorded on a smoked drumn by means of a pin attached to the rattle.

Klauber did extensive work on the characterisation and identification of rattlesnakes, and wrote *Their Habits, Life Histories, and Influence on Mankind*, which was published in 1956.[54] Clearly the antics of rattlesnakes at different temperatures was a subject which several people were intrigued by.

Even though Klauber's work was not as early as that of Shapley, it appears that several scientists were having, or had had, similar ideas. One of these was Amos Emerson Dolbear (1837–1910), whose work preceded that of the others by a couple of decades. In 1897 he published a paper entitled

Amos Dolbear, ca. 1880.
Image from the Historical Materials Collection (UA136), Digital Collections and Archives, Tufts University.
(Sourced from Picryl.)

The Cricket as a Thermometer, the title of which sums up his thoughts succinctly.[55] Could a cricket really act as a proxy thermometer? Clearly the work which came later showed that both ants and rattlesnakes could, but Dolbear chose the cricket. Like the ants, crickets are insects and therefore cold-blooded. Biologically there is no reason why they should not behave the same.

Dolbear's paper, by today's standards, is amazingly short, a mere 255 words[56] including formulae. It starts by describing the chirping of the crickets in an almost romantic style:

> An individual cricket chirps with no great regularity when by himself and the chirping is intermittent, especially in the day time. At night when great numbers are chirping the regularity is astonishing, for one may hear all the crickets in a field chirping synchronously, keeping time as if led by the wand of a conductor.

Dolbear quickly states (it was a very short paper, after all) that the chirping is dependent on the temperature, and says, "Thus at 60° F. the rate is 80 per minute." He continues by reporting that for every rise of one degree Fahrenheit in temperature the chirping rate of the animal increases by four chirps per minute. Interestingly, he finds that there is a lower limit to this, being about 50 °F:

> Below a temperature of 50° the cricket has no energy to waste in music and there would be but 40 chirps per minute.

50 °F equates to 10 °C. This lower limit is not far from that reported for the rattlesnakes. Shapley talked about a scale 30 °C long, with an upper limit of approximately 40 °C. It is therefore no great surprise that the data for a range of cold-blooded animals are similar, even across insects and reptiles.

Dolbear's paper then proposes an equation which can be used to link the rate of chirping of the cricket to the air temperature. This is then re-used to abruptly end the paper with a question:

> For example. What is the temperature when the concert of crickets is 100 per minute?

$$T. = 50 + (100 - 40)/4 = 65°$$

Here, the "100" is the speed of chirping and can be superimposed to enter the measured chirp rate at any given time. This formula has, not surprisingly, been called Dolbear's Law. Clearly, he had no doubt that the measurement of the rate of the cricket's chirping sound can be used to assess the ambient temperature, with no real thermometer needed. However, in this paper he didn't give any data points which he measured, and he didn't even give the species of animal he used. It is likely that he was using the snowy tree cricket, *Oecanthus fultoni*, but the formula can be used as an estimate with other cricket species too.

Amos Dolbear graduated from Ohio Wesleyan with a BA in 1866, and went on to take up the position of assistant professor of natural history at Kentucky University. However, he is best known for his work as a distinguished physicist. He is credited with being an inventor of early versions of the radio and the telephone.[57,58] Like Shapley, his work with insects was only part of his portfolio of scientific discoveries. Also, like Shapley, Dolbear's biological endeavours appeared to be outside his main line of work. However, unlike Shapley, it is the paper on crickets for which Dolbear is probably most well known.

Dolbear's work on crickets has been built on by others. His 1897 paper has been cited 72 times.[59] Like Shapley, it was used by Uvarov in 1931,[60] but has also been quoted when the synchrony of insect behaviour is discussed.[61,62] A study was extended to use field crickets,[63] and even in the 21st Century in papers reporting on neuronal function, Dolbear's work is still quoted.[64] Now in 2021, Dolbear's work, like that of Shapley, is still being used.[65,66] It even appears in popular culture, being mentioned by the 'scientists' in *The Big Bang Theory*, an immensely popular US television comedy series about three physicists and an engineer working at Cal Tech in Pasadena.[67] Sheldon Cooper – a theoretic physicist – was heard to say, when challenged about the chirping of a cricket:

> In 1890, Emile Dolbear determined that there was a fixed relationship between the number of chirps per minute of the snowy tree cricket and the ambient temperature; a precise relationship that is not present with ordinary field crickets.[68]

> [Note that the name is slightly wrong, as is the date, as Dolbear's paper was 1897.]

However, even Dolbear was not the first to recognise that biological activity and temperature are related. This accolade appears to go to Margarette W. Brooks, who published a report about the temperature relationship of crickets in 1881.[69] The report by Brooks has been pointed out by several people, as far back as 1899, for example, by Robert T. Edes, who published a note in *The American Naturalist*.[70] He was quoted[71] in saying:

> A few years ago a note appeared in the Boston Transcript calling attention to the very exact dependence of the rapidity of the chirps upon the temperature of the surrounding atmosphere and giving a formula therefore . . . possibly the same.

Margarette W. Brooks seems to have disappeared into relative obscurity, unlike Dolbear, who has been given all the credit. Although she apparently published other papers in *Popular Science Monthly*, little is known about her. She was a secretarial assistant to the zoologist Edward Morse (Margarette's sister Josephine was his housekeeper).[72] Morse had an interest in Japan and published on a range of Japanese artifacts. In 1885 he wrote:

> At the same time I desire to thank Miss Margarette W. Brooks for much aid given to me in my work.[73]

He died in 1925 at the age of 87, in Salem, Massachusetts, United States. There was a Margarette W. Brooks in Salem (1860–1929),[74] and it is tempting to assume that this is the same person.

However, even Brooks seemed to be using work which preceded her. She seemed to be testing something which she had read in "the Salem

Gazette, signing himself W.G.B.", but the exact identification of this person appears to still remain unknown. However, they are quoted in saying:

Take seventy-two as the number of strokes per minute at 60° temperature, and for every four strokes more add 1° and for every four strokes less deduct the same.

This is remarkable in its similarity to what Dolbear was saying 16 years later. However, Brooks did some experimenting, and unlike Dolbear's paper she gives data. She says: "After seeing this [W.G.B.'s work] I determined to make a number of observations, to find out if this were an invariable rule. I tried it." Brooks gave her observations from 12 evenings, from 30th September to 17th October, and even discussed the accuracy of her experiments. There was a "heavy frost", the crickets were hard to hear at times but her thermometer, which she states was "no standard thermometer", was more exposed to the elements than the crickets which were in the trees, and therefore sheltered from the wind. She presented the chirp rate, the temperature as calculated using W.G.B.'s equation and the temperature recorded on her thermometer. Despite her reservations about her experimental set up she concluded that there was a relationship between the chirps that she counted and the air temperature, saying there was a "remarkable accordance between the number of vibrations and the temperature of the air". However, she concluded, in a very modest way, that: "With more accurate observations doubtless a closer agreement would be proved." Considering that her last observation was on the 17th October, the paper was published in the 22nd October. Clearly, she did not sit on her data, unlike some scientists, such as Sir Isaac Newton who was notorious for delaying his publications.[75] As for Brooks, I have not seen any further work on crickets. She only seemed to have published one paper on this topic. Sadly, she seems to have had no credit for this work either.

Because Brooks was using work that preceded her, it has been suggested by some that the law should be renamed as the W.G.B. Law.[76] However, 'Law' itself seems to be a strong, and perhaps misleading naming. It has been pointed out that several factors other than air temperature can affect the chirp frequency of crickets, including wind currents and the physiological condition of the animals, amongst other things.[77] Perhaps crickets are not very reliable thermometers in real life after all.

Over the years there have been different variants on Dolbear's Law, so that it can be easily used. So, if you are out in the wilds and wish to know the temperature you can count, "the number of chirps in seven seconds and adding 46".[78] This was discussed by Clausen in 1957.[79] If you can't find a cricket, or hate the outdoors, you can build your own, and then use the electronic cricket – the Electron Cricket Sensor – to tell you the temperature using Dolbear's Law.[80] All you need to do is replace the battery, once it is built – it was supplied as a kit.

For a pragmatic way to use insect thermometers it is worth considering a hand drawn diagram in Eric Sloane's book.[81] This was reproduced in a Master of Science thesis written by Kristy Wallisch.[82] Here the diagram (Drawing 3 in the book) indicates – with a central thermometer – that insects are mute at 40 °F, and are dormant at lower temperatures, clearly noting the lower end of their usefulness. Between 55 °F and 76 °F Sloane helpfully suggests how the call of the katydid[83] changes, from "Kate!" through derivations of this to "Kay-tee did it!" at the higher temperatures. He also points out the males are voiceless, so this is all females singing. However, centrally down the diagram are the chirp rates of the cricket, ranging from 0 to 194, through the temperatures of 40 °F to 85 °F. On the right is a quick calculator for the user. For the black field cricket simply count the chirps for 14 seconds and add 40, and then you have air temperature in °F. To convert to °C use the formula: $(°F - 32) \times 5/9$, although Sloane did not give this conversion. What he did show was that at a little above 100 °F all insects become quiet or idle, including bees, which gives the top temperature to which this methodology can be employed.

2.6 Biochemical explanation – why animals move faster as the temperature rises

The simplified explanation of how animals can be a proxy thermometer comes from the nature of the biochemical reactions which can take place. As mentioned earlier, as the temperature increases, so does the rate of reaction. All organisms are dependent on efficient conversion of nutrients taken in into a useable energy source. The common energy molecule used by cells is adenosine triphosphate, which is usually referred to as ATP. This is not recreated from scratch when required, but formed by the simple phosphorylation (addition of a phosphate group) of ADP, that is, the diphosphate form. This creates in ATP what is often referred to as a 'high energy bond', which can be broken, recreating the ADP and releasing the energy at a site distant from ATP generation. This energy can then be used to drive muscles, for example.

To make ATP there is a central pillar of metabolism, made up of three main pathways: glycolysis, Krebs cycle[84] and oxidative phosphorylation. Ironically, glycolysis uses ATP at the early stages so instead of making ATP it uses it. However, in later stages glycolysis makes enough ATP to overcome this deficit, as well as make some extra, so making a profit, so to speak. This is important because some cells only use glycolysis as a source of ATP, and these are known as glycolytic cells. White blood cells, such as neutrophils, manage to generate enough ATP in this way to survive, as they do not really need a large source of energy until their final moments of glory when they destroy invading pathogens, and themselves in the process.

In the absence of oxygen cells may re-route their metabolism down an anaerobic pathway. In yeast this results in the production of alcohol (ethanol), so useful for beer and wine production. They also release carbon dioxide in the process, which is why they give us well-risen and fluffy bread. For most cells this is not the most efficient use of their nutrients and they need to take carbon sources – such as glucose – to full oxidation. To do this they use the Krebs cycle.[85] However, this does not make ATP. The Krebs cycle does generate a similar molecule called GTP (guanosine triphosphate), and many modern biochemistry books tend to draw in ATP at this point in the pathway, but it is not strictly the reaction which takes place. GTP is hugely important in cells, controlling a suite of proteins known as the G proteins, for example. These proteins are used for the control of a wide range of functions in cells, such as the response to adrenaline (epinephrine).

Even though the Krebs cycle makes no ATP per se, it is the main source of compounds which can be used to subsequently drive ATP production. The molecules made by the Krebs cycle are known as redox compounds, as they undergo reduction and oxidation. They are 'taxis' for electrons. These electrons are shuttled into the mitochondria and drive the electron transport chain (ETC). Peter Mitchell (1920–1992)[86] was instrumental in working out what was going on, but he was not taken seriously for quite some time. He eventually won the Nobel prize and went on to establish the world-famous Glynn Research Institute in Cornwall. What Mitchell suggested, which was so hard for the other biochemists to agree on at the time, was that the ETC pumps protons across the mitochondrial membrane and hence sets up an electrochemical potential (also known as protonmotive force[87]). This could then be used to drive ATP production through an ATP synthase enzyme (otherwise known as an ATPase). This was found to be correct – hence his Nobel Prize for Chemistry in 1978[88] – and is a common mechanism used in a range of biochemical processes, including photosynthesis. The same process was discussed earlier when we looked at thermogenesis.

The metabolic pathways do not run unchecked, but rather have enzymes which are rate limiting. These are control steps. However, not only will they be rate limited by the biochemical processes that regulate them, but they will have a maximum rate which may be limited by temperature. As the cells warm up in cold-blooded animals, these rate limiting steps, along with all the other enzymes involved, will be able to act faster, making the intermediate compounds leading to ATP production. Furthermore, the ATP is used in the muscles by an ATPase enzyme, which is very similar to the one found in mitochondria. All such enzymes will be temperature dependent, and therefore as the temperature rises so does the ability to make ATP and use it. Hence animals are more active.

Of course, there are limits to the temperature range of metabolic pathways. This is clearly seen in the data when people have tried to use animals as thermometers. Anything close to the freezing point of water is the lower limit, which is not a surprise when we remember that the reactions are mainly taking place in water. The upper limit is given as proteins start to struggle to maintain their structures, and hence functions, as temperatures rise above 40 °C. However, between these limits, the activity of some animals will be partly dependent on the ambient air temperature, and this is what is being observed in ants, snakes and crickets. However, as pointed out by others, the nutrient state, health and other environmental factors which impinge on an animal's physiology will have an impact, which means it is an even bigger surprise when Brooks, Harlow and others had such a straight correlation in their data.

2.7 Chapter summary

There seems to be no doubt that the measurement of the activity of cold-blooded animals can be used to estimate the temperature, and through time this has remained known as Dolbear's Law (for crickets), even though he was not the first to publish on this idea. It should be called Brooks's Law, or even W.G.B.'s Law. The relationship of temperature and animal behaviour appears to be amazingly accurate, and can hold true for a range of species, from ants and crickets to snakes. The underlying mechanism relies on the fact that all the animals used are cold-blooded, and the temperature has a direct effect on the metabolic rate of the cells of those animals. This makes perfect sense and is relatively easy to comprehend: the higher the temperature, the more energy available to those animals for running or shaking their tail. This would not work for warm-blooded animals where the cells' temperature is highly regulated.

The relationship between animal behaviour and temperature has a novelty factor, even getting into common culture and fun instrumentation, but it has little practical application in real life. However, such work is still used and does help to lead to a greater understanding of the physiology, cell biology and biochemistry of these biological systems.

These observations also show that animals can be very sensitive to their environment. Ants run faster, crickets chirp more quickly et cetera, so there is no reason that they cannot use this sensitivity to alter their behaviour too. In the next chapter, the use of animals to forecast the weather will be discussed, and it may be that an animal's sensitivity to temperature is part of a suite of perceptions used for animals to know if it is about to rain, or that the day will be hot and dry. After all, Shapley observed that if his experimental room was too warm the ants became "became erratic and sluggish". Perhaps the behaviour of Shapley's ants can not only tell us the temperature but signal to us that changes in the weather are on the way.

Endnotes

1 Hölldobler, B. and Wilson, E.O. (2010) *The Leafcutter Ants: Civilization by Instinct.* WW Norton & Company, New York.
2 Dawkins, R. (2021) *Flights of Fancy: Defying Gravity by Design and Evolution.* Head of Zeus Ltd., London. ISBN: 9781838937850
3 National Geographic: www.nationalgeographic.com/science/article/carboniferous (Accessed 02/11/21)
4 No publication could be found for this research.
5 Phys Org: https://phys.org/news/2019-12-leafcutter-ants-stormy-weather.html (Accessed 21/12/21)
6 IPCC Sixth Assessment Report: www.ipcc.ch/report/ar6/wg1/ (Accessed 02/11/21)
7 Al Jazeera: www.aljazeera.com/gallery/2021/2/8/in-pictures-himalayan-glacier-flash-floods (Accessed 10/12/11)
8 The BBC: www.bbc.co.uk/news/science-environment-58494641 (Accessed 02/11/21)
9 New York Times: www.nytimes.com/2020/08/17/climate/death-valley-hottest-temperature-on-earth.html (Accessed 02/11/21)
10 Evans, S.S., Repasky, E.A. and Fisher, D.T. (2015) Fever and the thermal regulation of immunity: The immune system feels the heat. *Nature Reviews Immunology*, 15, 335–349.
11 Time: https://time.com/6053214/thermometer-history/ (Accessed 02/11/21)
12 Brine is high concentrations of salt in water.
13 Physics World: https://physicsworld.com/a/in-praise-of-lord-kelvin/ (Accessed 02/11/21)
14 It is worth noting that an increase of 1 °C is the same as an increase of 1 °K, it is just that the scales start at a different point.
15 At 0 °K motion within molecules would stop, and therefore it is impossible to go below this temperature, and hence it is called absolute zero.
16 The lowest temperature recorded is just 38 picokelvins (pK) above absolute zero (38 trillionth of a degree): The Daily Mail: www.dailymail.co.uk/sciencetech/article-10072549/Scientists-create-coldest-temperature-recorded-lab.html (Accessed 02/11/21)
17 Britannica: www.britannica.com/biography/Thomas-Johann-Seebeck (Accessed 21/12/21)

18 Grodzinsky, E. and Levander, M.S. (2020) History of the thermometer. In *Understanding Fever and Body Temperature* (pp. 23–35). Palgrave Macmillan, Cham.

19 National Library of Medicine: https://medlineplus.gov/ency/article/001982.htm (Accessed 22/06/22)

20 Nicholls, D.G. and Locke, R.M. (1984) Thermogenic mechanisms in brown fat. *Physiological Reviews*, 64(1), 1–64.

21 Thermogenin is an uncoupling protein, as it 'uncouples' the movement of electrons from ATP production in the mitochondria of cells. Palou, A., Picó, C., Bonet, M.L. and Oliver, P. (1998) The uncoupling protein, thermogenin. *The International Journal of Biochemistry & Cell Biology*, 30(1), 7–11.

22 Many compounds are recycled by cells, and not made from scratch *de novo*. ATP/ADP cycles are typical examples, but others include the recycling of NADH to NAD^+ and NADPH to $NADP^+$. There are numerous other examples. This no doubts reflects the need to conserve as much as possible as most organisms are not resource replete, as discussed in the text.

23 The idea of surface area to volume ratios was also discussed by Dawkins in his book on the evolution of flight: Dawkins (2021).

24 Mogk, A., Ruger-Herreros, C. and Bukau, B. (2019) Cellular functions and mechanisms of action of small heat shock proteins. *Annual Review of Microbiology*, 73, 89–110.

25 Adders, or vipers, are the only venomous snakes in the UK.

26 Live Science: www.livescience.com/32585-how-do-lizards-cool-off-.html (Accessed 22/06/22)

27 Jim Clark: www.chemguide.co.uk/physical/basicrates/temperature.html (Accessed 22/06/22)

28 Heikkila, J.J. (2017) The expression and function of hsp30-like small heat shock protein genes in amphibians, birds, fish, and reptiles. *Comparative Biochemistry and Physiology Part A: Molecular & Integrative Physiology*, 203, 179–192.

29 For an excellent text on Shapley's work see: Sobel, D. (2017) *The Glass Universe: The Hidden History of the Women Who Took the Measure of the Stars*. 4th Estate, London. ISBN: 9780007548187

30 Sobel (2017), p. 196.

31 It should be noted that he also published an earlier paper which appears to be on ecology: Shapley, H. (1919) *The Bulletin of the Ecological Society of America* (vol. 3, p. 4). Ecological Society of America, Tucson, Arizona.

32 Shapley, H. (1920) Thermokinetics of *Liometopum apiculatum* Mayr. *Proceedings of the National Academy of Sciences of the United States of America*, 6(4), 204.

33 *A Bug's Life*, made by Pixar Animation Studios for Walt Disney Pictures in 1998: https://disney.fandom.com/wiki/A_Bug%27s_Life (Accessed 21/12/21)

34 *In vitro*, from the Latin for "in glass", meaning in an experiment in the lab where the samples have been isolated from the animal. *In vivo*, in the living animal.

35 Shapley, H. (1924) Note on the thermokinetics of Dolichoderine ants. *Proceedings of the National Academy of Sciences of the United States of America*, 10(10), 436.

36 Shapley, H. (1966) Oral history transcript – Dr. Harlow Shapley. (Niels Bohr Library and Archives, College Park, MD), quoted in Huey, R.B. and Kingsolver, J.G. (2011) Variation in universal temperature dependence of biological rates. *Proceedings of the National Academy of Sciences*, 108(26), 10377–10378.

37 Krogh, A. (1914) *Zeit. Allg. Physiol.*, 16, 163 and 178. Volkmann, Dissertation, Freiburg, i.B., 1908. (As quoted by Shapley).

38 Citation numbers taken from Google Scholar (October 2021).

39 Uvarov, B.P. (1931) Insects and climate. *Transactions of the Royal Entomological Society of London*, 79(pt. 1).

40 Cox, C.L., Tribble, H.O., Richardson, S., Chung, A.K., Curlis, J.D. and Logan, M.L. (2020) Thermal ecology and physiology of an elongate and semi-fossorial arthropod, the bark centipede. *Journal of Thermal Biology*, 94, 102755.

41 Hurlbert, A.H., Ballantyne, F. and Powell, S. (2008) Shaking a leg and hot to trot: The effects of body size and temperature on running speed in ants. *Ecological Entomology*, 33, 144–154.

42 Andrew, N.R., Hart, R.A., Jung, M.P., Hemmings, Z. and Terblanche, J.S. (2013) Can temperate insects take the heat? A case study of the physiological and behavioural responses in a common ant, *Iridomyrmex purpureus* (Formicidae), with potential climate change. *Journal of Insect Physiology*, 59(9), 870–880.

43 Levy, H.M., Sharon, N. and Koshland Jr, D.E. (1959) Purified muscle proteins and the walking rate of ants. *Proceedings of the National Academy of Sciences of the United States of America*, 45(6), 785.

44 Chick, L.D., Waters, J.S. and Diamond, S.E. (2021) Pedal to the metal: Cities power evolutionary divergence by accelerating metabolic rate and locomotor performance. *Evolutionary Applications*, 14(1), 36–52.

45 Lindenberg, A. (2021) *Timing of Sensory Preferences in Camponotus Ants* (Doctoral dissertation, Universität Würzburg).

46 The full interview can be found here: American Institute of Physics: www.aip.org/history-programs/niels-bohr-library/oral-histories/4888-2 (Accessed 22/06/22)

47 Shapley did publish another paper on ants, but not about temperature dependence: Shapley, H. (1920) Note on Pterergates in the Californian harvester ant. *Psyche*, 27(4), 72–74.

48 Not all cold-blooded animals run, such as snakes which have no legs.

49 Martin, J.H. and Bagby, R.M. (1972) Temperature-frequency relationship of the rattlesnake rattle. *Copeia*, 482–485.

50 4 °C is the approximate temperature of most domestic refrigerators.

51 Chadwick, L.E. and Rahn, H. (1954) Temperature dependence of rattling frequency in the rattlesnake, *Crotalus v. viridis*. *Science*, 119(3092), 442–443.

52 The action potential is the electrical activity of the nerves (neurons) which will trigger the movement of the muscles. For a review on the subject see: Barnett, M.W. and Larkman, P.M. (2007) The action potential. *Practical Neurology*, 7, 192–197.

53 Klauber, L.M. (1940) *Occasional Papers San Diego Society of Natural History*, No. 6, 13. Laurence, M. *A Statistical Study of the Rattlesnakes*. National History Museum, San Diego: https://www.ns66.sdnhm.org/science/publications/occasional-papers/ (Accessed 20/12/22)

54 More recent editions are still available: University of California Press: www.ucpress.edu/book/9780520040397/rattlesnakes (Accessed 02/11/21)

55 Dolbear, A.E. (1897) The cricket as a thermometer. *The American Naturalist*, 31(371), 970–971.

56 I determined this by transferring a copy to MS WORD, doing some quick editing for formatting marks and then doing a word count. I might have been inaccurate, but it is a close estimation.

57 Ohio Wesleyan University: www.owu.edu/files/resources/dolbear.pdf (Accessed 02/11/21)

58 Dolbear's new telephone system. *Science* (1881) July 2;2(54), 310–313: https://doi.org/10.1126/science.os-2.54.310

59 Data from Google Scholar (October 2021).

60 Uvarov (1931).

61 Buck, J.B. (1938) Synchronous rhythmic flashing of fireflies. *The Quarterly Review of Biology*, 13(3), 301–314.

62 Walker, T.J. (1969) Acoustic synchrony: Two mechanisms in the snowy tree cricket. *Science*, 166(3907), 891–894.

63 Pires, A. and Hoy, R.R. (1992) Temperature coupling in cricket acoustic communication. I. Field and laboratory studies of temperature effects on calling song production and recognition in *Gryllus firmus*. *Journal of Comparative Physiology. A, Sensory, Neural, and Behavioral Physiology*, 171(1), 69–78.

64 Robertson, R.M. and Tomas, G.A.M. (2012) Temperature and neuronal circuit function: Compensation, tuning and tolerance. *Current Opinion in Neurobiology*, 22, 724–734.

65 Leith, N.T., Macchiano, A., Moore, M.P. and Fowler-Finn, K.D. (2021) Temperature impacts all behavioral interactions during insect and arachnid reproduction. *Current Opinion in Insect Science*, 45, 106–114.

66 Banerjee, A., Egger, R. and Long, M.A. (2021) Using focal cooling to link neural dynamics and behavior. *Neuron*, 109, 2508–2518.

67 Season 3, Episode 2, *The Jiminy Conjecture*.

68 Quotes: www.quotes.net/mquote/886090 (Accessed 22/06/22)

69 Brooks, M.W. (1881) Influence of temperature on the chirp of the cricket. *Popular Science Monthly*, 20, 268.

70 Edes, R.T. (1898) Rate of the chirping of the tree cricket (*Oecanthus niveus*) to temperature. *American Naturalist*, 33, 935–938.

71 Berenbaum, M. (2008) Entomological bandwidth. *American Entomologist*, 54(4), 196–197.

72 Museums of the Bethel Historical Society Online Collections & Catalog: https://bethelhistorical.org/catalog/item/SERIAL_1.26.2.1 (Accessed 22/06/22)

73 Scribd: www.scribd.com/read/358357131/Japanese-Homes-and-Their-Surroundings (Accessed 22/06/22)

74 Wikidata: www.wikidata.org/wiki/Q18342462 (Accessed 02/11/21)

75 There has been some discussion about whether Newton's delay in publication was a myth: Derbes, D. (2010) A twenty year delay in Newton's publishing? *American Journal of Physics*, 78, 1077–1078.

76 Berenbaum (2008).

77 Frings, H. and M. Frings (1957) The effects of temperature on chirp-rate of male cone-headed grasshoppers. *Neoconocephalus ensiger*. *Journal of Experimental Zoology*, 134, 411–425.

78 Berenbaum (2008).

79 Clausen, L.W. (1954) *Insect Fact and Folklore*. The Macmillan Company, New York.

80 Ramsey Publication: https://cld.bz/bookdata/opfeELe/basic-html/page-1.html (Accessed 02/11/21)

81 Sloane, E. (2005) *Eric Sloane's Weather Book*. Courier Corporation. The text – first 20 pages which includes the relevant drawing on p6 – is publicly available at: https://books.google.co.uk/books?hl=en&lr=&id=6gg6Nri6jkUC&oi=fnd&pg=PP1&dq=Eric+Sloane%27s+weather+book&ots=mSmi95BC2-&sig=o9uppbqsbZbjmfrEiK7uNuo2WCc&redir_esc=y#v=onepage&q=Eric%20Sloane's%20weather%20book&f=false (Accessed 14/12/21)

82 Wallisch, K. (1999) *Animal Behavior as a Weather Predictor* (Doctoral dissertation, Oklahoma State University). Although Google Scholar cites this as a doctoral thesis, it was in fact submitted for a Master of Science.

83 A nocturnal insect otherwise known as the long-horned grasshopper or bushcricket.

84 Named after Hans Krebs (1900–1981). The Nobel Prize: www.nobelprize.org/prizes/medicine/1953/krebs/biographical/ (Accessed 30/06/22). There is a famous picture of Krebs on his motorbike, or 'cycle' which is often shown to students as a joke. It can be seen in an article by Whitehead, D. (2010) Oh to be in Oxford now that Krebs is there. . .!. *The Biochemist (Lond)*, 32, 54–55: https://portlandpress.com/biochemist/article/32/5/54/1887/Oh-to-be-in-Oxford-now-that-Krebs-is-there (Accessed 20/12/22)

85 Otherwise known as the citric acid cycle (CAC) or the TCA cycle (tricarboxylic acid cycle). They are all the same thing.

86 Although Mitchell was born in Mitcham, South London, he died in Bodmin, Cornwall, where he had established the Glynn Research Laboratories. These laboratories appear to have been taken over by University College London, establishing what they refer to as the Glynn Laboratory of Bioenergetics: www.ucl.ac.uk/~ucbt294/ (Accessed 03/07/22)

87 This mechanism is not restricted to eukaryotes, but is found in bacteria too: Maloney, P.C., Kashket, E.R. and Wilson, T.H. (1974) A protonmotive force drives ATP synthesis in bacteria. *Proceedings of the National Academy of Sciences, U.S.A.,* 71(10), 3896–3900.

88 Nobel Prize: www.nobelprize.org/prizes/chemistry/1978/summary/ (Accessed 03/07/22)

3

Predicting the weather

3.1   Introductory story

It had been a long struggle against the weather. They had lost more days than he had anticipated, and the Commission was starting to get annoyed with the lack of progress. He had promised that the harbour wall would be finished before the end of September and now it was the end of October, and it was likely that the site would need to be abandoned for a third winter. Another delay could cripple the whole project and he was hoping to be given more work after this contract. He needed it. He had a growing reputation as an engineer but not delivering on his promises was not going to help. He needed this harbour to be finished, and quickly.

"Crane over the next one," Arthur instructed. The waves were lapping at the base of the wall, but there was nothing to be alarmed about. They should be able to get three more hours of work done before it was too dark to see. The men were used to long hours, which made up for the days of idleness as the storms had come and then diminished.

"We'll set the corner next," the foreman shouted. "Not that one, the one next to it."

The men lashed the rope around the massive granite block and, as Arthur watched, it was hoisted over the gap between the barge and wall. He looked on, extremely satisfied as the stone slid neatly into the space that had been left for it. His design and measurements remained precise, and he knew that this wall would stand up to the raging seas for decades to come. He would be remembered as the person who had constructed a harbour where everyone had said it was impossible to do so.

"And the next. Get the one on top and we can work back towards the cliff," Arthur shouted to the foreman.

"Right'o, sir," the foreman shouted back.

"Excuse me, Mr Jenson."

DOI: 10.1201/9781003343844-3

"Yes, boy, what do you want?"

"I think a storm is coming," the young lad said. He was one of the junior workers who had been drafted in. His clothes were ripped and dirty, and he looked as though he hadn't washed for a week. He had plucked up all his courage to approach the boss, and was shaking inside.

"And why do you think that?"

"I've been watching the fish."

"You've what? I pay you to work, not stare at the fish."

"But they've disappeared. Normally there are lots on the reef, but they've all gone. They've gone to survive the next storm. It happens every time."

"Rubbish, boy. Get back to work, or you'll be seeking employment elsewhere by the end of the day. I hear that the workhouse is looking for scruffs like you."

"Yes, sir." The young boy ran back to his position and started to lash the next stone ready to be lifted into place. He wished he'd never said anything. He certainly hadn't obtained the boss's respect. He would keep his head down in the future.

In the middle of the night, Arthur Jenson was standing at the window of his room. The rain was lashing at the glass and he was struggling to see. However, as the lightning lit the sky, it clearly illuminated that the end of the harbour wall was missing. There had been no time for the mortar to set and the day's work now lay at the bottom of the sea. The granite stones had been quarried, dressed and transported eighty miles before being tossed away by the waves. It was a major setback and would mean work would now be stopped until April. Fish! Perhaps he should start watching the fish too? It would have saved a considerable loss and embarrassment. The young lad he would sack in the morning. It only seemed fair.

3.2   The weather and animals – an introduction

Throughout the last 2,000 years people have had sayings about watching animals and predicting the weather. The *Farmer's Almanac*[1] has phrases such as: "The early arrival of crickets on the hearth means an early winter." This purports to be able to predict the weather for the next few months, saying what the whole season will be like. Others, such as noting the cows are sitting down, will predict the imminent arrival of rain, with a much shorter timescale of forecasting.

Although nearly always viewed with scepticism, such observations seem to have some pragmatic use. In 1914, Sir Ernest Shackleton set out on the ill-fated Imperial Trans-Antarctic Expedition. On 18th January 1915, during this epic adventure his ship, *Endurance*,[2] was trapped in sea ice. On 21st November 1915, the ship was eventually crushed and sank. The crew had drifted north, living on the ship while it was trapped in the ice of the Weddell Sea, but eventually they had to abandon the ship on 27th October 1915, and camp on the ice. The Antarctic weather was predictably cold but also storms were an issue, threatening to either blow away the crew's vital equipment or to break up the ice flows. As part of their survival strategy Shackleton turned to the animals, although there was not a plethora of them to watch. Nonetheless, it was purported that Shackleton watched the birds, both to predict if bad weather was arriving and whether the weather would make the icepack unpredictable.[3] It was their only 'land' after all. The men could not launch the small boats they had rescued from *Endurance* as they would be crushed by the ice, but they needed enough ice to camp on. On occasion the ice would crack, and the men had to jump across to more secure ice to survive. Therefore, any indication that danger was imminent was vital to their survival and observing the behaviour of birds helped.

And survive they did. Shackleton made an epic final voyage back to South Georgia Island in one of the small boats, the *James Caird*, as they had named it. They had left on 5th December 1914 and arrived back on 10th May 1916. A boat was then sent to rescue the rest of the original crew, who had been left on Elephant Island, approximately 650 nautical miles away. It is interesting to note that watching birds did not only give Shackleton

Postage stamp from South Georgia depicting Shackleton's ship *Endurance*, beset in the Weddall Sea, circa 1972.
(Sourced from Shutterstock. Contributed by Sergey Goryachev, but made black and white.)

an indication of the weather, but also whether they were approaching land. Cormorants rarely fly more than 15 miles from land, so on seeing such birds the *Endurance*'s crew could celebrate that they were within reaching distance of solid land.

Observing animals to forecast the weather has featured in popular fiction too. A good example of this is in the book *The Cove*, by Alice Clark-Platts, published in 2022.[4] One of the characters in the story, Harry, notices that the trees are full of birds. He then points out that they are on the leeward side and suggests that the birds are huddling together on the side of the trees which is farthest from the wind, and hence the approaching storm. Noah notes that he fails to "smell the approach of rain", and remains sceptical of the value of this observation. Meanwhile Simon teases Harry by calling him "the bird whisperer". There is no mention of the species of bird they are supposedly observing, and of course a major storm then hits the island, adding to the drama of the story.

With adventurers and fiction writers both turning to animals to predict the weather, perhaps less scepticism is warranted. Numerous examples are discussed later in this chapter and listed in Table 1. However, two main questions need addressing. Firstly, is there any validity to these stories? Can animals really be helpful, and if so on what timescale? Secondly, if animals can be used to predict the weather, how are they doing it? What is it that they are sensing? The answer to this final question is sought in the discussions that follow, and summarised in Table 2.

3.3 Why watch the weather?

Watching and predicting the weather can certainly save lives and infrastructure. Whether it is the ravaging of storms, or the heat of summer, known to buckle railway lines,[5] it is useful to know what the environment is about to throw at us.

Especially in the UK, there seems to be an obsession with the weather. There hardly seems to be a day that goes by without someone mentioning it. As reported in the *Independent* newspaper,[6] in a study[7] commissioned by the Business Development Director of Bristol Airport, Nigel Scott:

> The average British person spends the equivalent of four and a half months of their life talking about the weather, a study suggests.

However, the forecasts that we receive are not always as accurate as they could be. One famous occasion in the UK was on 5th October 1987. The BBC had a well-known weather forecaster called Michael Fish. He was seen on national television saying: "Earlier on today, apparently, a woman rang the BBC and said she heard there was a hurricane on the way. Well, if you're watching, don't worry, there isn't!"[8] He went on to say that the winds would be farther south, across Europe. At the time I was living in a university hall of residence in Bristol. Later that day we had trees lying across the entrance road, trees that had stood for decades. The whole of the south of the UK was hit and there was major disruption in what has been dubbed the 'Great Storm'.[9] The winds reached 115 mph, and approximately 15 million trees were brought down. Sadly, 18 people died. It just goes to show that despite all the instruments and technology available to the weather forecasters in 1987, a woman who was probably observing her own way of predicting the weather was the one who was correct.

Weather forecasting was started by Admiral Robert FitzRoy in the mid-19th Century.[10] He appears to have coined the word 'forecasting' and he started the Meteorological Office in the UK (abbreviated to Met Office by most who refer to it). He is better known as the captain of *HMS Beagle*, the ship on which Charles Darwin sailed around the world in the 1830s and led to the publication of *The Origin of Species*[11] in 1859, changing the thinking around evolution forever, and still relevant today. As for the Met Office, it is also still instrumental today in predicting the weather, and gives short- and long-range forecasts around the world.[12]

Weather forecasting, disseminated by newspapers, television news and now prominently on the internet, remain instrumental to the way people live. Such information remains vital for protecting people and infrastructure. Near Bristol (UK) there is a famous motorway bridge – the Severn Bridge.[13] It was said at the time of it being built in the 1960s that this bridge had a contour akin to an airplane wing and hence would be safe from high winds. However, strong winds regularly blow up the Severn River and the authorities which manage the roads need to monitor the bridge and close it when necessary. Partial closures are triggered by winds of 40 knots (46mph), with total closure if the wind gusts of 60 knots (69mph). Even the new bridge (Prince of Wales Bridge opened in 1996) closes if wind speed exceeds 70 knots (80mph).[14] This is a powerful illustration of how watching the weather has consequences even if no damage is occurring. If both bridges are closed it is a long detour north to reach the next major crossing over the river.

There are attempts to mitigate the effects of high winds. Windsocks[15] are often seen at airports[16] and allow an estimation of side winds which may make landing a plane or helicopter difficult. They are similarly seen on high road bridges, and anyone who has seen a lorry overturn will appreciate why.[17] Whilst in Switzerland a few years ago I was driving in horrendous weather and a lorry in front of me rose upon its wheels on one side to an angle of about 45 degrees, but fortunately did not topple completely over. It was a scary thing to see, but highlighted the force that the wind can bring to bear.

On a larger scale the world has seen devastating weather in many places recently. Flooding in Germany appeared to be hard to predict[18] but had terrible consequences.[19] Similar events were reported from China,[20] and as I write this there is a major flood on the news in Chennai, India, with 14 so far reported to have been killed.[21] On the other hand, many areas have suffered from the lack of water, and drought conditions are as devastating.[22] Even major rivers, such as the Colorado River, have drastically reduced water levels, with consequences across national borders.[23]

The Intergovernmental Panel on Climate Change (IPCC), which is a United Nations based group which assesses the science underpinning climate change, raised major concerns in their report in 2021.[24] In the summary for policy makers, they state:

> It is unequivocal that human influence has warmed the atmosphere, ocean and land. Widespread and rapid changes in the atmosphere, ocean, cryosphere and biosphere have occurred.

They go on to say: "Each of the last four decades has been successively warmer than any decade that preceded it since 1850." This is likely to have significant consequences for large regions of the world. The IPCC reports also says: "Human influence has likely increased the chance of compound extreme events since the 1950s. This includes increases in the frequency of concurrent heatwaves and droughts on the global scale." It has been predicted that regions of the world may become uninhabitable. In a recent paper, Rosemary Wakeman[25] states:

> Mass migrations – particularly by communities along coastal areas and floodplains as whole regions become uninhabitable – are already upending social stability. No person, no community will be able to withstand the impact of floods, droughts, fires, and storms without competent and active government assistance.

In another paper by Lokendra Sharma, the survival of island states was discussed and, rather alarmingly, it is stated:

> Even if the islands are not completely submerged due to rising sea levels, they may become uninhabitable due to several factors like damage to the coral reef, intrusion of saltwater in the freshwater table, rising food insecurity and extreme weather events. These

threats question the notions of territoriality, sovereignty and state-hood of island states in a post-submergence/uninhabitability scenario, challenging the very basis of the current international system of nation-states.

Some island nations, such as the Maldives in the Indian Ocean,[26] or the Marshall Islands[27] in the Pacific Ocean, may simply cease to exist in the future if sea levels rise too much. If that happens, the people will be displaced, and their cultures will no doubt be lost forever. The ecosystems for animals and plants will also disappear, with potential for permanent species loss. Once gone, these islands will be lost in their present state, and even if they re-emerge in the future, inevitably they will be different.

3.4 Instrumentation which has been developed to measure the weather

Over the course of history there has been a range of tools to measure the weather, including thermometers (as discussed in Chapter 2), and the barometer. These are discussed later in this chapter, and further listed in Table 2.

Today, supercomputers are used to try to work out what may be coming in the near future. An example is the use of the Roshydromet supercomputer technology.[28] Even so, people still moan about the weather forecast being wrong and the accuracy of what is given to the public is questioned.[29] Animals have a place here too, with a recent children's website suggesting that eight animal groups are good indicators. These are: frogs, birds, cows, insects (ants and others listed separately), sheep and elephants.[30]

Weather forecasting is a very complex and expensive undertaking and involves the measurement of many aspects of the environment, including temperature, air pressure, precipitation, wind speed and wind direction, as well as the position of the jet stream.[31] Air pressure is immensely important to the prediction of weather patterns. This can be used to predict the arrival of low- or high-pressure regions of the atmosphere, often with associated cold and warm fronts. This can then be used to predict rainfall, as well as temperatures. In the UK, deep low pressures are associated with storms, usually approaching from the west across the Atlantic. The Met Office now names these storms.[32] At the time of writing, the last storm was Barra, named on the 5th December, two days before it made landfall across Ireland and the UK. The next would be Corrie (they are named alphabetically).

The measurement of air pressure is through the use of a barometer. This instrument comes in different types, although such an instrument has been used for a very long time. In 1944, a review of the history of the barometer stated[33]:

> It is a remarkable fact that the most common means of measuring atmospheric pressure for meteorological purposes is still that devised by Torricelli. There is no fundamental difference between the 'Torricellian tube' and the modern 'fixed cistern' barometer, the latter being purely a mechanical improvement on the original instrument.

A Torricellian tube is an evacuated glass tube which is filled with mercury. Originally the open end of the tube would have been placed in a pool of mercury and then the height of the mercury column in the tube would be used to measure the atmospheric pressure. Obviously having open mercury is very unwise, given that it is extremely toxic.[34] However, the use of mercury was not always frowned on. Augustin Fresnel, in 1825, suggested that a vat of mercury could be used in lighthouses, as a lubricant:

> I propose to float our rotating devices, of the first order, in a bath of mercury, instead of placing them on rollers. This project won't present many difficulties; nevertheless, as I have not put it into execution, I won't require you to adopt it for your first lighthouse.[35]

> [It was adopted by many lighthouses[36]]

For a barometer the tube of mercury has to be long enough, as the atmospheric pressure can raise, in a vacuum, a column of approximately 76 centimetres at sea level, so the glass tube needs to be longer than this. The original idea was suggested by Evangelista Torricelli (1608–1647) in 1643, whilst he was in Pisa, Italy.

When I was at school, making a mercury barometer was a regular demonstration in physics lessons. A metre-long glass tube was simply filled with mercury and carefully inverted, and the open end placed under the level of a pool of mercury in a glass dish. The mercury would appear to miraculously drop in the tube, leaving a vacuum above the mercury's meniscus, and the air pressure holding up the mercurial column. It was quite magical to see. On one occasion the teacher wanted us all to see the procedure well and held up the glass tube. He poured in the mercury which went straight through the bottom of the tube, shattering the glass and spreading mercury all over the floor.[37] Now we know how toxic mercury is, it is no surprise that such practices are banned.[38]

Variations of this mercury-in-a-tube type barometer have been made since, with modifications being suggested by Robert Hooke (1635–1703),[39] who used a U-shaped tube and also added a wheel scale. Others also have had their ideas to bring to the problem. These included a diagonal barometer[40] and a balance barometer described by Sir Samuel Morland (1625–1695).[41] Another to have a go at improvements was the French scientist Guillaume Amontons (1663–1705), who tried to make a barometer for use at sea.[42] Later, William Derham, Rector of Upminster (1657–1735),[43] used a rack and pinion mechanism, and Stephen Gray[44] included a microscope for reading the measurements.

However, not all barometers are based on mercury. An aneroid barometer relies on a sealed metal chamber. This then expands and contracts as the atmospheric pressure changes. This was invented in 1844 by a French scientist from Nantes, Lucien Vidi (1805–1866). This is the basis of many domestic barometers hanging in the lounges and halls of many houses. Many of these have a dial which even tries to predict the weather, saying "Stormy", "Rain", "Change", "Fair" or "Very Dry", respectively as the air pressure increases. More recently,[45] there has been the invention of digital barometers and readings are more accurate and easier to see. They are also easier to carry.

Other important instruments for weather forecasting include the rain gauge[46] and anemometer (*anemos* is Greek for wind). The former will indicate how much precipitation has fallen, as it is often a simple catchment of water with a scale (snow fall needs a different measure or often just a

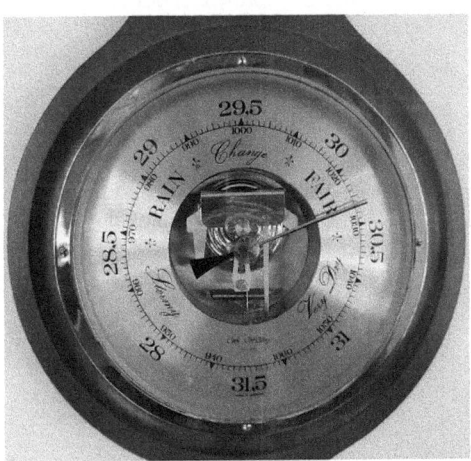

A typical household barometer showing how it predicts the weather, depending on where the hand points.
(Photograph by the author. This image has been made black and white.)

US Weather Bureau snow gauge: 1909.
Photo from National Photo Company Collection at the Library of Congress.
(Sourced from Picryl.)

measuring stick). There are variations on rain gauges, including ones which are based on optics or acoustics. An anemometer for measuring wind speed is more complex than a simple version of the rain gauge and involves a set of (usually three) cups which rotate on a central vertical axle. The faster the 'rotor' spins the higher the wind speed, but there is no indication of direction, which is usually given by a windvane (or weathervane). The windvane was said to have been invented by a Greek astronomer, Andronicus, in 48 B.C.,[47] so they have been around for a long time. However, modern ultrasound anemometers will give wind direction as well as speed, as discussed later.

The anemometer is a more recent invention than the windvane.[48] The oldest version of an anemometer was probably created by Leon Battista Alberti (1404–1472) in 1450. However, Robert Hooke is also credited

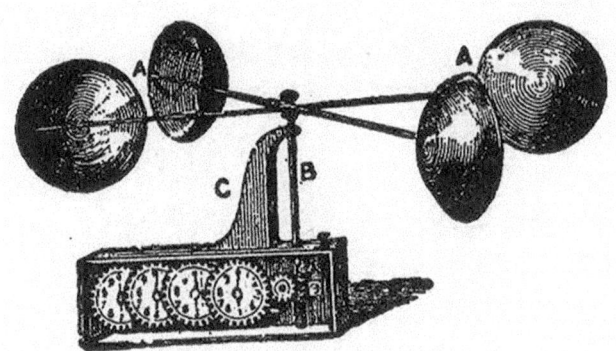

Anemometer: From Student's Reference Book, Chicago, F. E. Compton and Company: 1914.
(Sourced from Picryl. This image has been made black and white.)

with inventing a similar version in 1664. A German known as Wolfius (Christian Wolff) is also credited with making a version in 1708.

What we might consider to be a modern anemometer was invented in 1846 by an Irish scientist called John Thomas Romney Robinson. This instrument is often referred to as a hemispherical cup anemometer. Robinson's version had four such cups rotating horizontally in the wind, but often modern anemometers have three cups. However, even though these are commonly seen atop weather stations around the world, their technology has been superseded by the sonic anemometer, invented by Andreas Pflitsch in 1994. Using sound bounced between two transducers, both the speed and direction of the wind can be measured. Such ultra-sonic anemometers are readily available.[49] From the operational information given by the manufacturers such instruments should be able to work at wind speeds of up to 65m/s (145 mph), with 2% accuracy, and work down to low wind speeds of 0.01 m/s. The sensor also gives direction to the nearest 1 degree and can be heated for working at low temperatures, having a range of −55 to +70 degrees C. And of course, these days, all the information can be put on social media sites for all to see.

Atmospheric humidity is a good indicator of weather. Humidity is measured with a hygrometer. The first hygrometer was thought to be invented by Leonardo da Vinci in the 1400s,[50] although the measure of humidity (or water content) of materials was probably carried out during the Shang Dynasty in China (or Yin Dynasty, approximately 1556 to 1046 BC). Several other people have improved the hygrometer, including Francesco Folli (1624–1685) in 1664. He called his instrument *mostra umidaria* and in 1665, he presented it to Grand Duke Ferdinand II de' Medici

A typical hygrometer.
(Photograph by Guy J. Sagi from Shutterstock, but made black and white.)

(1610–1670), who it appears liked what he saw.[51] Other people who worked on the hygrometer were Horace-Bénédict de Saussure (1740–1799)[52] in 1783, who also invented an anemometer, Robert Hooke (1635–1703) and John Frederic Daniell (1790–1845) in 1820.

Many of the hygrometer designs rely on the movement of a material as it absorbs water, and the higher the humidity the more water there is to absorb. Hooke's instrument used the husk of oat grain, which curled and so the mechanical movement could be measured. Others were based on hair, which often curls when it absorbs moisture, such as developed by Horace-Bénédict de Saussure. Today, *Accuweather* even gives a forecast of how curly hair may be depending on the weather forecast,[53] a reverse-engineered form of the hair-hygrometer. Of course, it works the other way too. If hair curls, it suggests that the air is humid, and perhaps rain is on the way, or has recently passed if the ground is hot and steaming.

A popular version of the hygrometer was a dew-point instrument. This was invented by Grand Duke Ferdinand II de' Medici in the mid-1600s. This is often based on the condensation of water onto a metal surface.[54] A variation of this is the psychrometer.[55] This has two thermometers, one which is kept wet, and the temperature difference recorded can be used to calculate the air's humidity. Modern hygrometers rely on the properties of semiconduction materials. As the humidity changes so does the material's electrical conductance, and hence a measure of the humidity can be shown.

Commercial weather stations, therefore, can measure the current weather with a range of modern instruments – many of which are listed in Table 3. However, home weather stations can now be purchased cheaply. For example, there is the BRESSER system.[56] Such systems give several measures of weather parameters, including temperature, humidity, rain fall, wind speed and direction (including gusts), UV and light levels, and air pressure. It also gives forecasts and allows results to be published on the internet for others to see, for example at Automatic WEather KArten System (AWEKAS:[57]), Weathercloud[58] and Weather Underground.[59] Such internet sites pull in information from a wide range of amateur weather stations and can then analyse the data to give accurate local information.

The weather can also be predicted using modelling tools.[60] This is augmented by views of the world from the satellites[61] which look down at us too. Weather prediction is a vastly expensive and technologically driven industry. The question raised here is 'can all this be replaced with an animal?'

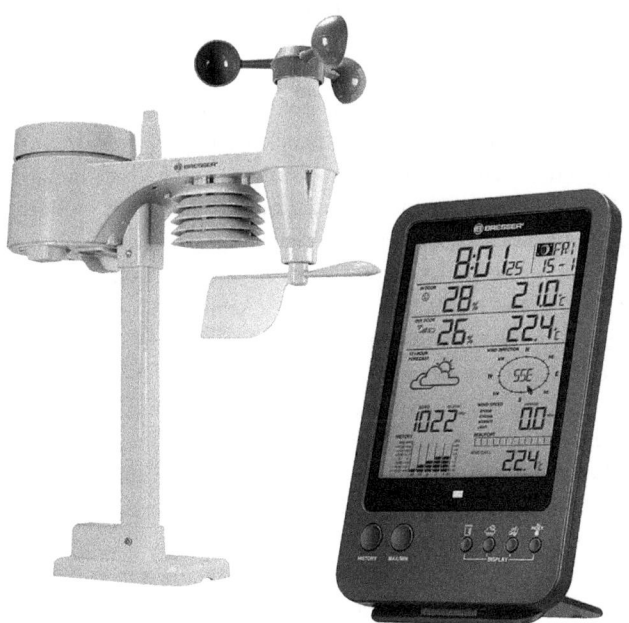

An example of a commercial home weather station. An external device sends the information to a digital screen. Both the anemometer and weathervane can be easily seen.
(Photograph supplied by and copyright to Bresser GmbH.)

3.5   Animals and how they can be used to predict the weather

The human obsession with the weather can be summed up by the introductory paragraph of a recent paper by Elizabeth Goldbaum[62]:

> It's seven o'clock on a gray Wednesday morning. As you stretch and yawn yourself awake, you reach out to grab your phone and open a weather forecast that's filled with sun, cloud, rain, or snow icons. These brightly colored images help you decide if you should wear extra layers, put on rain boots, or pack your sunglasses as you start planning your day.

Weather, and climate changes, are therefore at the forefront of concern around the world, and as such, were discussed at the 26th UN Climate Change Conference of the Parties (COP26) in Glasgow, UK[63] at the beginning of November 2021. Therefore, monitoring and predicting the weather around the world is a topic which will only increase in relevance in the future. Even though forecasting the weather has been attempted since the Greek philosophers such as Aristotle,[64] and today we have numerous instruments to aid us, including those discussed earlier, there may still be a place for the study of animal behaviours to predict weather changes.

The construction of the Bell Rock lighthouse was thought to be an impossible task. The Bell Rock is a sandstone reef which lies approximately 27 miles east of Dundee and 11 miles south of Arbroath, right in the middle of the North Sea shipping lanes. For centuries it was a hazard to ships and many tragedies unfolded there. Originally known as Inchcape Rock, in the 14th Century the *Abbot of Aberbrothock* (Arbroath) fixed a bell on the rock to warn sailors but it was removed by pirates.[65] However, the name stuck and ever since it has been referred to as the Bell Rock. In the beginning of the 19th Century, after a long battle with parliament, a commission was granted to John Rennie[66] (1761–1821) to build a permanent lighthouse there, but the design and work was effectively given to Robert Stevenson (1772–1850), grandfather of Robert Louis Stevenson (1850–1894), the latter who wrote such works as *Treasure Island* and *Dr Jekyll and Mr Hyde*. Robert senior had already built up a reputation for creating lighthouses on land, but this was a quantum leap in difficulty. The Bell Rock was underwater at high tide, and at low tide was a treacherous, slippery rock covered in seaweed. The beginning of the project was the construction of a temporary beacon tower which would also be the accommodation for the workmen. In the meantime, they had to live on a ship called the

Smeaton, after the builder of the Eddystone lighthouse,[67] John Smeaton[68] (1724–1792). The men then had to carve out of the Bell Rock a foundation for the new lighthouse, which was being constructed in stone, copying to a large degree the advances made by Smeaton at the Eddystone, namely the use of interlocking stones to stop the sea moving them. This could only be carried out at low tide, as the water would cover their workings as the tide rose. The workmen were working at sea whilst there, living on a boat and then in a temporary beacon on the rock. Weather, therefore, was always on their minds. A large wave could easily wash them off the rock, and storms would make it impossible to land supplies, including rations, and would also potentially endanger their lives if their accommodation was washed away. After all, at the Eddystone in 1703 (on 27th November), the first lighthouse was completely washed away with the loss of all hands, including the architect. It had been built, in a rather flamboyant style, by Henry Winstanley[69] (1644–1703). The tower completely disappeared in the storm, so those working at the Bell Rock would have known the type of danger they were in. As can be seen from a later painting by Turner,[70] the seas at the Bell Rock could be mountainous, reaching to the top of the lighthouse.

In the beacon on the Bell Rock, Stevenson had his own small cabin, enabling him to sleep separately and have some privacy from the men. In this "he kept a cot-bed, a folding table, a few books, a barometer and two small stools."[71] Apparently, he said that his only essential item there was his Bible, rather surprisingly, not his barometer. However, despite the weather being so important, and the fact that he had a barometer to predict storms, he also watched the animals as indicators of possible impending doom. He was:

> noting the habits of the sea birds and the way in which shoals of fish clustered over the reef in good weather but vanished into the deep as bad weather approached.[72]

With the weather being so important to Stevenson's survival, and that of his men, he turned to animals to help him predict what might be happening in the future. It was too late if the storm hit and they were still working on the reef. It was better to stop work for the day and retire quickly to the beacon and wait to see what happened, even if the storm never arrived.

The forecasting of weather using animals is certainly not a new idea, even for the early 19th Century. In a work entitled *De signis* by Theophrastus, there are these quotes about animals:[73]

> It is a sign of a long winter when sheep or goats have a second breeding season.

And:

> The behaviour of the hedgehog is also significant: this animal makes two holes wherever he lives, one towards the north, the other towards the south: now whichever hole he blocks up, it indicates wind from that quarter, and, if he closes both, it indicates violent wind.

Theophrastus was from Eresos in Greece, and was born in 371 BC. He was a pupil and successor of Aristotle. He obviously thought the observations of animals was worth writing about. He says: "centipedes in large numbers as a sign of rain" and "Swallows striking the surface of a lake with their bellies signal rain". At one point, Theophrastus tries to explain why animal behaviour may make sense:

> likewise cattle and birds fighting for food more than usual [are a sign of storm], since [γάρ] they are getting prepared in advance.[74]

This, of course, makes sense from the animals' point of view. If bad weather is coming, then stocking up with food may be a good survival strategy. Likewise, an animal such as a hedgehog protecting its den from possible disaster is a good thing to do. A sensible temporary measure such as this may be all that is needed to survive the storm. However, some of the observations just seem bizarre and would be hard to explain: "A seal making its loud sound in the harbor while holding an octopus is a sign of storm."[75] How this action of a seal can be related to rain is anyone's guess. Besides, how often would this happen? Perhaps this was observed before a major storm and the myth persisted. After all, as we will see later, the arrival of a storm may be able to be predicted by animals by their sensing of air pressure, so unusual behaviour is not that uncommon. However, it is hard to imagine how watching for this particular observation would be useful.

The work of Theophrastus and others was picked up in a 1952 paper by D.W. Ewer,[76] from the University of Natal, including the indications given by gnats, fleas and hedgehogs. If fleas bite then rain is imminent, which is also predicted if "cocks, hens or hawks are searching for lice". Ewer says that as well as hedgehogs, badgers and field mice block their holes against the wind, indicating the direction from which the wind will blow. This makes sense from an animal point of view as it would give them protection. However, it does mean that such animals can not only predict the wind increasing in intensity, but the direction from which it will come.

Greek philosophers were not the only ones to be looking to the animal world to explain and predict the weather. Many cultures around the world have myths and anecdotes where animals are used to determine what the weather may be like in the near future. In what has been termed indigenous knowledge systems,[77] C. Makwara Enock,[78] focussing on Africa, discussed many examples where people have observed animals and suggested rain and other weather events. This is summed up:

> The results showed that agricultural activities were related to wind systems, migratory and mating trends of wild animals and the position of the moon. The colour of the horizon at sunrise and sunset, appearances of rare animals and bird breeding patterns in river valleys were used in drought and flood prediction so were the flowering patterns of certain plant species.

The rather long paper uses several examples, spanning many animal species – these are collected together in a table. The behaviour of insects is highlighted as being of particular importance. For example: "When a lot of crickets are observed on the ground, a poor rainy season is expected." However, another species of insect seems to predict the opposite: "By contrast, when jerrymanglums/sun spiders (dvatsvatsva) are visible in the area, they signal the imminent arrival of a wet spell." This is obviously crucial information if crops are to be planted or harvested. Just as Shapley observed, ants are also useful.

> Heavy rains are predicted when ants emerge from their holes in large numbers to collect food from homes and the veld and this is associated with an impending long wet spell. The ants disappear less than twenty-four hours before the storm. Ant behaviour triggers farmers to collect firewood to dry places in preparation for a long wet spell.

Rain can also be assumed if ants build defences around their nests, or are seen in unusual places, such as on the ceilings of rooms. This would make sense as the insects are future proofing their existence from what they perceive as a looming threat.

Higher organisms are also worth observing. Enock reports: "The appearance of black and white stork (shohori/shuramvura), denderas, swallows (nyenganyenga) are associated with a good season and eminent rain." The direct observation of the animal may not be necessary either, rather, if it can be heard then indications of the weather may also be possible. Enock also says:

> The singing of some such birds is said to be a good omen in so far as rainfall is concerned. In particular, if dendera birds are heard

singing, it is believed to be a very good sign of an approaching good rainy season.

Furthermore: "If a cuckoo bird is heard signing, rains are said to be just about to fall," whilst the presence and croaking of frogs is also an indicator of rain. Additionally:

> The local type of kingfisher is associated with heavy falls within days of its appearance as the interpretation is that the nature of the sound that it produces resembles clattering of rain drops characteristic of a heavy downpour.

These animals appear to be signalling rain, but not doing much about it. It seems as though they are letting others know about the rain. It may not be perceived as a risk but perhaps a good thing, for example, if it has been dry and water is relatively scarce. Rain is not always something which needs to be avoided. Especially for amphibians such as frogs, the forecast of water may be something to celebrate, rather than dread.

Some of the predictions seems to be quite long-range forecasting too, not predicting rain later that day, or even tomorrow, but forecasting that the whole season may be good, or bad. Storks and swallows seem to be important for this.

> Equally, if a lot of swallows and white stork are seen in the locality, they are indicative of the onset of a promising rainy season. Indeed, a stork flying at a very high altitude is associated with a good season.

Birds seem to be good indicators of weather:

> drought in Muroyi area is anticipated when waterfowls (masekwe and hurekure) breed on the ground and in lower patches on flood plains.

Again, this makes sense. If the birds perceive that there is not going to be a lot of rain, then they would feel safe breeding in places where water may have otherwise gathered. On the other hand, if they are wrong, it could be disastrous as their nests would get washed away.

Enock also gives a good example in Australia, where hawks have been used to predict the ending of a drought. If the birds sit together on a tree branch and any latecomers to the party cannot land, then it is said that the drought has reached its 'lowest point'. This seems an amazingly inexact way to predict anything, as it depends on the branch and number of birds.

The activities of mammals are worth observing too. Enock states: "When game animals give birth in large numbers, it signifies a normal to above normal season and the reverse is true." However, regardless of the species, many of these animal behaviours discussed in Enock's paper make sense.[79] Animals either seem to be trying to get away from impending doom, by going to higher ground if rains are coming, or preparing for a good season, as food is likely to be plentiful. However, what is not discussed here is *how do the animals know?* Flying animals, such as butterflies and birds, may have an inkling from direct observations, but how would an ant know? But despite the lack of a knowledge of *how* animals are forecasting the weather to come, the observation of such activities is still useful for those who need to know whether it will rain or not. Furthermore, such myths and anecdotes have been embedded in human cultures and have been used for a long time. Clearly, there is something in these observations which lasts the test of time and is worth passing down through the generations.

Others too have researched the use of animals in different cultures to forecast the weather. Risiro *et al.*,[80] with a focus on the Chimanimani District of Manicaland, Zimbabwe, also tabulate many examples. Rain is imminent if there is singing of storks, or if there is singing and flying of *haya* birds. If spiders are running around, or swallows fly at low altitude, rain is going to be very soon. The rainy season is predicted by the presence of millipedes or frogs, or the breeding of goats. Once again ants are useful here, as if they are searching for food or they are sealing their nests, the rainy season is on the way. On the other hand, the hot season is predicted by presence of a large number of reptiles, the singing of insects (such as *nyenze*) or "Rock rabbit crying in the morning and evening". It is worth noting that many of these examples are very similar to those reported by others, so there is clearly a pattern here.

Reporting on East Africa, Radeny *et al.*[81] also list several examples of animals being useful. As before, there is a range of animals being observed. Imminent rain was suggested to alter the air pressure, so changing the animals' behaviour. The authors also try to gather evidence of how commonly such myths are used. For example:

> In Uganda, croaking of frogs during the day (as cited 68% of households in Rakai and 43% in Hoima) and singing by some specific birds are widely used to predict likelihood of rainfall onset.

Other indicators of rain include the emergence and croaking of frogs, as well as: "Ducks stretching their wings and playing in the dust is a sign of rainfall onset, especially the short rains."[82] The same behaviour of chickens also indicated rain was coming. In Lushoto, Tanzania:

> Occurrence of large flocks of swallows and swans roaming from the South to North during the months of September–November, for example, is an indication of onset of short rain sometimes within two to three days.

The activity of mammals was also important. If during the dry season animals such as monkeys, leopards, antelope and baboons came into the village it was a predictor of the coming season having good rains. In Borana, the digging of holes by the ground squirrel is indicative of a normal rainy season. Interestingly, "a high birth rate of male children and animals in a particular season is usually taken as a sign of a likely drought season." The reason for this is hard to imagine. It is known that ambient temperature may determine the sex of the offspring in many animals such as reptiles. A 1 °C change in temperature could make the crucial difference in the determination of sex ratios. This has been thought to be a possible problem with future climate change.[83] However, most animals, including mammals, would not have the sex of their offspring determined by ambient temperatures.

Of course, insects were also prominent in weather forecasting.[84] Rain was predicted by the presence of army ants, and the movement of black butterflies from the south was a good omen that rain was coming. Furthermore: "Presence of insects on Albizia trees with water dripping from them is an indication of a good season." Large numbers of bees and locusts suggested that there would be long rains, whilst the occurrence of flying ants (*Odentotermes* spp.) indicated there would be enough rain to allow planting of crops. This, as noted earlier, could be vital information for a remote village where the people depend on reliable crops for survival.

In Uganda, a survey of which weather forecasting observations were used was undertaken.[85] Of relevance here, the authors said that there were only a few common and frequently used indicators of the coming of the dry season:

> These included the appearance of bush crickets (*Ruspolia baileyi* Otte), winds blowing from the east to the west, the appearance and movement of migratory birds such as cattle egrets (*Bubulcus ibis* Linnaeus), and calling by the Bateleur eagle (*Terathopius ecaudatus* Lesson).

For the onset of the rainy season the indicators included: "cuckoo birds (*Cuculiformes: Cuculidae*) start to call, and winged African termite (*Coptotermes formosanus* Shiraki) swarms leave their nests."

Birds appear to be a good indicator of changes in both weather and climate in New Zealand. Māori knowledge of forecasting from Aotearoa[86] has been collected,[87] and there are many examples of where animals are used. If the falcon (Kārearea) screams on a fine day, it will rain on the next day. Conversely, if they scream on a rainy day, the next will be fine. Similarly, if the morepork (Ruru) screams more than once at night, rain is on the way. A storm and heavy rain are predicted if the swamp hen (Pūkeko) heads for high ground, or the native parrot (Kākā) is seen twisting or squawking above the trees. Probably the most bizarre indicator involved fish. If stones are found in the stomach of the Rāwaru, it is said that bad weather is coming. It is hard to think how the fish could possibly predict this, and why they would eat stones, unless they are trying to alter their buoyancy so they can migrate to deeper water. Fish rely on the use of a swim bladder for buoyancy.[88] Perhaps eating stones can supplement the lack of buoyancy when needed, so fish can retreat to deeper water which is below the level of surface disturbances. On the other hand, fish are known to suck algae and microbes off stones, so they may be looking for a quick snack.[89]

Longer, climate changes are also indicated by many Māori birds. The arrival of shining cuckoos (Pīpīwharauroa) or godwits (Kūaka) signals that a warm season is on the way. The squawking of the bittern (Matuku-hūrepo) suggests that a season of floods will come, whilst a season of gales and heavy snow will follow the presence of many heron (Kōtuku) in the summer.

In Australia, the Aboriginal culture not only uses animals to predict the weather, but animals are said to control it. In a 2011 paper by Prober *et al.*[90] there is a statement which says:

> for the Karajarri of northwestern Australia, the rain birds *kitirr* (Fork-tailed Swift, *Apus pacifus* Latham, 1802) and *wiyurr* (Barn Swallow, *Hirundo rustica* Linnaeus, 1758) were seen not only to indicate the imminent occurrence of rain, but to 'pull in the rain'.

It is very hard to see how animals can directly influence the weather. Seeding clouds, and the creation of rain, is possible. It was reported that in China, 1,110 rockets were fired into the sky to ensure that the 2008 Beijing Olympics were dry on the opening day.[91] The rain was made to fall before the ceremony was to begin. However, for less frivolous reasons cloud seeding has been used or proposed elsewhere, such as in Australia,[92] where rain may be needed for maintaining crops. However, how birds could have such a substantial effect on rain creation is very hard to imagine.

The activities of animals were also said to predict the activity of other animals: "For example, in southeast Queensland, string-like processions of hairy caterpillars predicted the clustering of breeding mullet in the waterways." Animals also indicated when certain plants were ready to harvest: "in British Columbia, the call of the Swainson's thrush (*Catharis (Hylocichla) ustulata* Nuttal, 1840) indicated that salmonberries (*Rubus spectabilis* Pursh) would soon be ripe and ready to harvest."[93] Another example being given by Lantz *et al.*:

> On the west coast of Canada, the Nuu-Chah-Nulth peoples of Vancouver Island recognize the correspondence between the ripening of the salmonberries (*Rubus spectabilis* Pursh) and the return of adult sockeye salmon (*Oncorhynchus keta* Walbaum) to freshwater.[94]

However, it is probably that the plant here is predicting (or altering) the behaviour of the animal, rather than the other way around. Clearly, though, this was important information about when food sources would be available or when certain agricultural practices should be undertaken.

Closer to home (for me, that is), in the UK there are also many examples of animals being used to forecast the weather. The recording of data from the environment may have started in the 18th Century. According to Lawrence[95]:

> the story begins in 1736 when Robert Marsham wrote 'Indications of Spring'. During 62 years, he recorded 27 natural events and information on 20 animals and plants in Norfolk.

One of the most common animal-based weather predictions is surely the observation that cows sit down when rain is coming. Today, this may be commented on in a flippant manner, but amazingly the publication *Science Focus* reported: "According to a recent survey by the UK Met Office, over 60 per cent of the British public believe that cows lying down is a sure sign of rain."[96] Ewer states: "It is widely known that cattle lying down in a field is a sign of rain."[97]

Cows may sense a change in moisture in the air, or a change in air pressure, with the suggestion that they are then trying to keep their future fodder dry by sitting on it. On the other hand, studies on thermal maintenance by cattle suggest that in sunny weather they would stand,[98] perhaps to try to keep cool. Perhaps the opposite is true when the weather is cooling off, as when rain is imminent. This has been suggested in a recent newspaper article,[99] although there is very little robust evidence for this.

Cattle-based weather predictions are not just found in the UK, so perhaps there is more to this. In their paper focused on East Africa, Radeny *et al.*[100] state:

> Drought indicators related to behavior of cattle include calmness and sleeping very close to one another in the pen, refusing to go and

graze in nearby pasture, preference for staying near watering points, loss of appetite for grass and salt, lack of interest in reproductive activities, and isolation from the rest of the herd for bulls. Other drought related indicators for cattle include weight loss with standing hair on skin, defecating and urinating in a sitting position, and with unusually reduced amounts of cow dung. For a normal rainy season, indicators related to cattle body condition and behavior include licking of each other's body, display of a relaxed mood and getting away from water points after drinking, normal reproductive behavior (including increased activity in bulls). In the absence of sickness, these body conditions and behaviors reflect future weather conditions.

Others suggest that cows might have different aberrant behaviour before a storm, including lying on their right side, licking their feet, sniffing the air, acting playfully or having lower milk.[101] Again, Ewer has some thoughts on this:

> Licking of the forehoof is a sign of rain; if they lick their left flanks or lie on their left sides the weather will be fine; if the bull leads the herd to pasture it will rain, but if 'he be careless and let them go at random' fine weather may be expected.[102]

But as pointed out in a 2013 paper:

> signs, such as watching the behavior of a herd of cows, require patience and are not stand-alone thereby requiring a combination of two or more other signs to be able to make an accurate weather forecast.[103]

There are many other weather indicators in the myths of Britain. Gulls seem to arrive inland looking for shelter. In a report about the southeast of England it is said that seagulls[104] arrive four to five days before the onset of bad weather.[105]

There are several other well-known cases of using animals to predict the weather. A good example is gnats.[106] Gnats flying in a particular manner seems to be an indicator of good weather. This I was regularly told as a child, as mentioned in Chapter 1. This behaviour was picked up by Ewer[107] in their 1952 paper, which starts:

> 'GNATS seem to be more worthy of esteem that the ordinary sort of almanac makers' wrote Thomas Mouffet at the beginning of the 17th Century, 'for they will tell you the weather at all

times, and for nothing, and that more certainly and timely. For if the gnats near sunset do play up and down in the open air, they presage heat; if in the shade, warm and mild showers; but if they altogether sting those that pass them by, then expect cold weather and very much rain.'

Thomas Mouffet (Moffett, amongst other variations of the spelling of his name) (1552–1604) was a physician who was born and lived in London before moving to Wiltshire. He obtained a BA in 1573, a MA in 1576 and a MD in Basel in 1578. He died in 1604. He wrote *Theater of Insects* which was published in 1634.[108],[109] This was published posthumously. Interestingly, it contained excellent sketches of insects, which were drawn before the improvements of the microscope latter in the 17th Century.[110] One of the pieces of advice that it is also reported that he gave was about keeping gnats at bay, for example: one should build "a 'Fen-canopy' out of hardened cow dung."[111]

"New Discoveries in Pneumatics." Humphry Davy (1778–1829) lecturing at the Royal Institution. Print by James Gillray.
(Available from the Wellcome Collection, CC BY 4.0 licence. This image has been made black and white.)

There is some sense in this as the insects will be carried by warming air currents as they rise in the atmosphere. This was apparently observed by Sir Humphry Davy,[112] as noted by others:

> As early as 1828, the famed scientist Sir Humphrey [sic] Davy, in his book Salmonia, had identified the feeding patterns of swallows as being central to this particular piece of popular wisdom: 'Swallows follow the flies and gnats,' he wrote, 'and flies and gnats usually delight in warm strata of air.'[113]

Of course, the adverse is true: as the air currents cool, they will tend to keep the insects lower to the ground, and this would be when rain may be on the way. It has been noted that several birds fly high when the weather is going to be fine, and this was explained by Davy's comments, as the birds will fly higher to feed on the insects. This would be particularly noted with swifts, martins and gulls, and would account for the ditty: "Swallows high – staying dry."[114]

As mentioned earlier, Ewer[115] also comments on a saying about the flea:

> 'When eager bites the thirsty flea
> Clouds and rain you sure will see.'

This is, perhaps, not an obvious advantageous change in behaviour, but maybe, like the leaf-cutting ants, the fleas sense that times of scarcity may be on the way and increase their feeding to compensate in advance.

Ewer also mentions some interesting wise words of others:

> Gilbert White records how his tortoise 'if attended to, becomes an excellent weather glass; for as sure as it walks elate, and, as it were, on tiptoe, feeding with great earnestness in the morning, so sure will it rain before night'.

And:

> Also Bartholomaeus Anglicus tells us that there was 'one in Constantinople[116] that had an urchin, and know, and warned thereby that the winds should come,

Bartholomaeus Anglicus[117] was a Parisian scholar, although of English descent. He compiled the encyclopaedia *De proprietatibus rerum* (On the properties of things). Although it is not clear when he was born, it was thought that the encyclopaedia was written around 1242 to 1247. It is thought that he died in 1272. The encyclopaedia itself was in 19 chapters, which included: 11. *De aere* (On the air), which included weather; 12. *De avibus* (On birds); 18. *De animalibus* (On animals).[118] Clearly, he had ideas and observations which have lasted the test of time.

The previous quote is probably from Reverend Gilbert White (1720–1793). A graduate of Oriel College, Oxford, he went on to write *The Natural History and Antiquities of Selborne*, which was published by his brother Benjamin in 1789. Richard Mabey, a naturalist and biographer, said: "Gilbert White's book, more than any other, has shaped our everyday view of the relations between humans and nature."[119] Again, the tortoise in the saying is feeding more before rain. Like for the ants, foraging would

Isabella tiger moth and caterpillar stage. Note the lighter (orange in reality) banding on the caterpillar. From *Field Book of Insects*, 1918.
(Sourced from Picryl. This image has been made black and white.)

be more difficult in wet and windy weather, so this would make sense for a wild tortoise. Just because a tortoise is kept in captivity does not mean that it would automatically lose its innate behavioural traits.

An observation of an animal and how it predicts the weather may not come from a behaviour pattern, but from the growth characteristics of the animal. It has been noted that some caterpillars have a different banding pattern depending on how bad the forthcoming winter may be.[120] The caterpillar stage of the Isabella tiger moth *(Pyrrharctia isabella, otherwise known as Isia isabella)* is known as the woolly worm, or woolly bear caterpillar. They are segmented (13 segments) and the central segments are rust-coloured giving the appearance of a band across their back. There is a superb little drawing of a woolly bear in Eric Sloane's book (drawing 2).[121] If this band is wider than usual, then this is an indicator of a mild winter.[122] If the band is narrow, then a bad winter is on the way. If there is no rust at all you are looking at the wrong species of insect, so there is no need to panic. Apparently, this observation began in the autumn of 1948 when Dr C. H. Curran started to collect specimens in Bear Mountain State Park, north of New York, USA. He did this for eight years. He then reported on what he thought the weather would be like in *The New York Herald Tribune*. Even though there seems to be quite a culture built up around this caterpillar-based weather forecasting there is little scientific evidence, and it has been suggested that the rust colour is indicative of the age of the caterpillar and is more a measure of what the last winter was like, rather what the next one will bring.[123]

Even so, some people take the woolly worm myth even further.[124] If the head is particularly dark the winter will start very cold, whilst a cold end of the season is predicted if the tail is dark. Alternately, if the animal is dark-brown central coloured at each end the bad weather will be in the middle of winter.[125] If the hairs are unusually long on the caterpillar, then the winter will be unusually cold, with the converse being noted. Lastly, if the 'worms' are heading south they are trying to escape a bad winter, and if they are heading north, they are prepared to stay as the weather will be warmer. How far a caterpillar could walk to escape the weather does not seem to be noted. Although there is quite a following for this little caterpillar, there seems to be little solid evidence that it does predict the weather, which is a pity, I feel.

It is interesting to note that these caterpillars have an ability to survive surprisingly low temperatures.[126] In one study, caterpillars were acclimatised to cold temperatures and then:

> all caterpillars could survive 1 week of continuous freezing at $-20°C$ or seven cycles of freezing–thawing at $-20°C$, but none survived freezing at $-80°C$.[127]

This is quite extraordinary for an animal. −20 °C is the temperature of a domestic freezer. It is no surprise that the caterpillars could not survive −80 °C, which is used in many science laboratories for long-term storage of biological materials, such as proteins and DNA. However, many animals, including insects and arachnids, do have antifreeze molecules,[128] including proteins,[129] so allowing them to survive cold temperatures which would make the average human stay indoors. Antifreeze mechanisms are often studied in fish, for example.[130] Therefore, the fact that woolly worms exist in the cold is not a surprise, they simply find the warmest place to nest. Even so, −20 °C is very cold. Having said that, cell cultures in the laboratory can be stored at liquid nitrogen temperatures,[131] as long as they are very slowly frozen and then thawed appropriately. Avoiding the formation of ice crystals is very important. Therefore, it is possible for animal cells to be frozen and still survive. It is still amazing for a whole animal. As any mountaineer or Antarctic explorer will attest, frostbite is not good and can lead to the loss of digits. For example, Sir Ranulph Fiennes lost the fingers on his left hand through frostbite after attempting a lone expedition to the North Pole.[132] Human tissue does not freeze and thaw well. Clearly the humble woolly worms have a trick that we could learn from.

Although there is a relative dearth of scientific evidence for how animals sense and respond to the weather this has not stopped some from writing Masters dissertations[133,134] on the topic. The latter is worth pausing over. It was written in 1999, by Kristy Wallisch of the University of Oklahoma, that she "examines the possibility that animal behavior can be used to indicate future weather". What she was particularly interested in was the feeding behaviour of birds before a storm. As it was Master of Science thesis, she had an element of primary data collection included. She carried out two studies, at Santa Fe, New Mexico, and Tulsa, Oklahoma, and looked at how much food was taken by birds from feeders. However, neither of her observations indicated any relationship between the food taken and the arrival of winter storms. Although she was no doubt disappointed in this (scientists tend to like positive results, which are usually easier to publish), she did say that her one study should not be used to dismiss the use of animals for weather prediction. What was great about the thesis, however, was the background and literature searching that she did to support her hypothesis. She drew on the work of others including a paper by Garriott written in 1903.[135] Here, the flight of birds in relation to air pressure was discussed and it was suggested that high pressure allows better flight whilst low pressure forces the birds to work harder to sustain flight, and hence they fly lower. If similar effects are seen with insects, which could be assumed, then the birds would be able to follow their food source too, as Davy had suggested earlier.

By cartoonist Clifford Berryman, this is entitled, "Two Prophets", one being a groundhog. The cartoon was originally published in the *Washington Evening Star* on July 18, 1907, but is now in the Berryman Political Cartoon Collection. It was drawn following the floods in Western Pennsylvania, Eastern Ohio and West Virginia.
(Sourced from Picryl.)

Several interesting examples of weather prediction were given too. Wallisch suggests that in the USA one of the best examples is Groundhog Day: 2nd February (Candlemas Day). She says:

A clear day on Candlemas produces the shadow that frightens the groundhog (*Marmota monax*) back into its hole for six more weeks of winter.

On the other hand, cloud would indicate that there is a low-pressure system dominating and an early spring may be on the way.

A second example involves goats (*Capra* sp.). If they were grazing high up on Mount Nebo, Oregon, it was a sign of good weather on the way. This idea was apparently picked up by a local radio personality who looked at the goats' behaviour during May 1971. Wallisch goes on to say:

> When the goats' behavior predicted the weather accurately 90% of the time, goat weather forecasts began and a 'Goat Observation Corps' formed. The goats ended their careers in the early 1990s when they were declared a traffic hazard and retired to a nearby farm.

Clearly observing the aberrant behaviour of some animals so that we may predict future weather can have its drawbacks.

In sizable tables, Wallisch's thesis lists a myriad of examples of animals being used for weather forecasting. These are arranged in different ways (either under animal observed, or weather predicted). Therefore, under rain there are a very long list of antics, including that of ants (Shapley would have been pleased!), bats, birds and mammals. Many of these are not unusual, with animals moving to lower land (e.g., sheep), or protecting their nests (ants). Some behaviours appear strange, such as dogs eating grass. This was also noted by Ewer.[136] I remember my pet dog doing this when I was younger, but I always assumed it was because of a digestive issue, or that they just like the taste – perhaps I should have taken more notice and run for a raincoat! For snow, birds were seen to feed more (which makes sense if they are preparing themselves for times of scarcity), turkeys (*Meleagris* sp.) perching in trees, cats facing away from fire and hornets nesting higher. Under storm there are a wide range of observations, including birds flying inland, sea urchins adhering to rocks (which makes sense), fish and marine mammals swimming upriver, deer and elk stampeding and mules laying back their ears. Lastly, some good news, perhaps (at least in the UK), and the prediction of sun, which includes birds singing and flying higher, goats moving to high land and dolphins splashing. Longer term forecasting is also listed, so an early spring will be coming if the flying squirrel calls in the middle of winter, for example.

What this extensive list of examples shows is that there is a large body of evidence, albeit mainly anecdotal and lore, that animals sense and react to future weather arriving. From Aristotle to now, this is a topic which has been used in a wide range of cultures, from all over the world. As Wallisch says in her conclusion:

> It is certainly feasible that all traditional lore on the subject is inaccurate, but the results of thousands of years of observation

and tradition from cultures from around the world should not be accepted or rejected on the basis of one study.

[she was referring to her work own here]

It seems a reasonable statement. All those myths and folklore have endured and no doubt have some use, even if some examples are both bizarre and hard to explain, or even to test empirically, as Wallisch discovered.

One of the fascinating discussions by Wallisch is the use of animals as barometers, and hence weather changes. Apparently frogs and leeches have been kept in jars of water, and their behaviour used to assess air pressure. Frogs croak for impending rain, or climb a little ladder and stay silent if fair weather is coming. However, as Wallisch explains, this idea was taken to extremes by George Merryweather, a man with a very suitable name.

George Merryweather (1794–1870) was a doctor in Whitby, Yorkshire, England. He was inspired by a poem written by Edward Jenner (1749–1823). Jenner was best known for his work on vaccines, in particular the smallpox vaccine.[137] Having noted that those who had previously contracted cowpox had a natural immunity to smallpox, Jenner realised that this could be used as a vaccine regime, and hence the first vaccination was born. Today, vaccination is thought to be the best way out of the COVID-19[138] pandemic.[139] Therefore, Jenner was a brilliant scientist, but what is not widely known is that he also published work on cuckoos[140] and wrote poetry, the latter as we noted was also undertaken by Sir Humphry Davy. It seems that some brilliant scientists also liked to write verse. In one of Jenner's poems, entitled *Signs of Rain*, is the line: "The leech disturbed is newly risen; Quite to the summit of his prison." It is thought that this inspired George Merryweather to create his leech-based barometer.

Ewer noted that the behaviour of the leech was mentioned in the Girl Guides' Diary. In his 1952 paper[141] he writes:

> Possibly simplest of all is to follow the advice given in a Girl Guides' Diary and keep a leech in a bottle. Then if good weather is coming, the leech will remain curled at the bottom of the bottle. If change-able weather is at hand, it will slowly move to the top and remain there until the weather is settled. If windy weather is due, it will keep moving rapidly through the water and will continue to do so until the wind has come.

This idea was taken by Merryweather and he created what was to become known as the Tempest Prognosticator.[142] This instrument contained 12 bottles arranged in a circle. Inside each was a single leech and a small quantity of rainwater. When the weather was fine the leeches stayed

TEMPEST PROGNOSTICATOR.

Merryweather's illustration of the Tempest Prognosticator.
(From the introduction to his essay; sourced from.[143] Also it is available from the Wellcome Collection under a Public Domain Mark. This image has been made black and white.)

at the bottom, but if rain was coming the leeches would work their way up the water and up the sides of the bottle. In the neck of each vessel was a small whalebone, and as the leeches wiggled into that space the bone was displaced and this caused a bell to be rung. Merryweather noted that some leeches appeared to be "stupid" as it was best if more than one managed to ring the bell. But it did seem to work and was reported to have foreseen a major storm which hit Whitby in October 1850, more than two days before it arrived.

The Tempest Prognosticator was an elaborate and impressive looking instrument, which stood about a metre high, and was made of mahogany, silver and brass, as well as glass. Merryweather was apparently also concerned for the welfare of the animals and designed it: "in order that the leeches might see one another and not endure the affliction of solitary confinement." He had great plans for the use of his instrument. He called it an "Atmospheric Electromagnetic Telegraph, conducted by Animal Instinct."[144] He exhibited it at the 1851 Great Exhibition[145] and tried to persuade the authorities to link one to the bell in St Paul's Cathedral in London. He was hoping to have lots made and used throughout the country, but they were not adopted and today if you wish to know what one looks like a good place would be Whitby Museum[146] where there is a replica on display. Probably the maintenance, and the thought of forever replacing the leeches was not overly appealing. Merryweather wrote and presented "An Essay: Explanatory of the Tempest Prognosticator in the Building of the Great Exhibition for the Works of Industry of all Nations." This was read before the Whitby Philosophical Society in 27th February 1851. It is available at the Wellcome Collection.[147]

Despite the rather weird nature and elaborate construction of this barometer, the Tempest Prognosticator does show that animals do appear to respond to barometric pressure, and can even be used as part of a forecasting instrument. However, I can't see this being adopted in the 21st Century and this instrument will be resigned to interesting history!

Returning once again to Wallisch, she cites Sattler,[148] who pointed out that animals have a constant need to assess their environments to survive, so it is not a surprise that some of the behaviours seen are related to weather and oncoming changes to environmental conditions. I obviously have no idea what mark was given for her thesis, but I enjoyed reading it – it is fully publicly available [see Footnote 101] – and I am sure it would have been well received as part of her degree.

The Tempest Prognosticator was not the only pragmatic way to use animals to predict the weather. In Ewer's paper[149] he also suggests the use of shark oil. He says:

> Possibly simpler then is to extract the oil from the brain and liver of a shark. If this is kept in a sealed bottle it will remain clear so long as the weather is to be fine, but if a storm is approaching it becomes cloudy, more particularly on the side from which the storm is approaching.

It is interesting that this is in a sealed bottle, so one assumes it cannot be affected by humidity. Also interesting is the point about it giving the direction from which the storm comes. In a newsletter from 1972[150] almost the exact words are used as by Ewer, but here the author is a E. De la Garza. They go on to say: "An increase in sediment in the bottle gives three or four days' warning ahead of time if a storm is in store." They report that

"Old-time fishermen" relied on this method of weather forecasting and that "I really value my shark oil weather forecaster, believe me, it works a lot better than my bad knee." Interestingly, they refer to the instrument as containing "barometric oil", and although they say that this is now written about in both scientific and popular literature, they do not cite any. The Bermuda Institute of Ocean Sciences (BIOS)[151] says[152] that the use of shark oil is local tradition and has been used for about 300 years. In 1985, *Yachting Magazine* featured an article on the use of a shark oil barometer.[153] They say that the sediment at the bottom of the jar is key. It forms peaks and troughs which are not static, and the highest peak indicates the direction from which the storm will approach. It also says that a person called Thatcher Adams had been studying the mechanism of this "barometer" and has concluded that it is due to the infrared radiation altering the squalene in the oil. Adams was actually a draftsman, but obviously had an obsession with these barometers. He seemed to have several which he observed on a regular basis. When someone played a prank on him by shaking one up, he became so excited that running into a shop he tripped and broke his leg.[154]

The article in *Yachting Magazine*, if anyone fancies trying it and has access to a dead shark, also gives quite a lot of detail of how to extract the oil, what type and age of shark to use, how to make the barometer, where to mount it (on the southwest corner) and how to observe it. It also suggests that the extraction of the oil is somewhat a smelly business and that your neighbours might not appreciate it. However, if you can manage to make this very simple device you can predict the onset of the next storm. And, apparently, they get better with age and some are said to be decades old, so you only need to ever make one. Any extras would certainly make a novel and interesting present at Christmas or for a birthday. It is quite interesting that the article ends by suggesting that you buy an electric barometer (needs an "AA" battery) instead, as it only cost $435 (in 1985, so not cheap) and was free of odour.

3.6 Biochemical explanation – what animals may be sensing to predict the weather

Animals have a variety of senses that may be used to determine what their environment is like. In humans we have touch, smell, taste, sight and hearing. However, in us many of these senses are quite limited and other animals may be able to sense things that we cannot. So, what are the senses used by animals to predict, for example, the arrival of a storm?

Interestingly, in the fiction book, *The Cove*,[155] one of the characters, Noah, comments that he fails to "smell the approach of rain". What would he be expecting to smell, considering that water has no smell? Certainly, we can smell the effects that rain may have, as it disturbs the foliage and ground around us. Perhaps Noah was expecting such smells to be wafted in on the wind? In fact, the smell of rain has a name: petrichor. This term was coined in 1964, referring to the golden liquid which acts as blood in the immortals.[156] The chemical involved is called geosmin,[157] produced by soil-living bacteria, perhaps as a way to ward off predators. It is reported that this highly pungent molecule can be smelt before rain, as well as after the rain. However, why it should be released before the water disturbs it in the soil is not discussed.

Mammals are not the only animals which can smell. As discussed by Ackerman[158] odour may have a role in the navigation of birds. She points to an experiment where the olfactory nerves of pigeons were severed before they were released in an unfamiliar location, with the expectation that they should return home, but they never did. Birds with unsevered, and therefore intact, nerves returned to their lofts. This suggests at least that these nerves are useful for a bird's navigation. As they approach home, birds can pick up smells which guide them. Ackerman suggests that nearly every species of bird has some ability to smell. They may use this for sensing food sources, for example, or to sense the presence of a predator. It is not inconceivable that such perceptions of smell are part of the suite of senses used to predict the changes in the weather.

However, there are more likely explanations of what the animals are sensing. In Clark-Platts's fiction book,[159] one of the characters points out that the birds are in the leeward side of the trees protecting themselves from the oncoming rain, and that birds are doing this because they can hear infrasound and that they are sensitive to changes in barometric pressure. Here, it is suggested that using such senses the birds are anticipating the oncoming storm, which indeed then adds to the storyline of the book. However, interestingly she has pointed out two possible mechanisms: sensing infrasound or barometric pressure.

There is evidence that the birds do have an inbuilt barometer and that they can sense the changes in air pressure which indicate that poor weather is coming. It is also suggested that they fly around in circles to "reset" this barometer. Although others suggest that gulls can sense low pulses of sound which may precede the arrival of a storm,[160] there is no reason to consider that there is only one sense being used here. It is not only in the popular press that such statements are made either. In a 2011 paper by Radhouani *et al.*, it is said:

> Birds are sentinel species whose plight serves as a barometer of ecosystem health and alert system for detecting global environmental ills.

However, in this case this is very misleading, because it is not the animals' behaviour which is being discussed but rather their health, and this paper is about the environmental pressure on antibiotics,[161] not the predicting of environmental conditions such as the weather. On the other hand, a study using the white-crowned sparrow (*Zonotrichia leucophrys*) concluded that:

> These data suggest that white-crowned sparrows can sense and respond to declining barometric pressure, and we propose that such an ability may be common in wild vertebrates, especially small ones for whom individual storms can be life-threatening events.[162]

Here, birds were placed in pressure chambers and then several parameters assessed, including behaviour and metabolic measurements, having taken blood samples after the pressure treatment. Control birds were kept at a constant pressure whilst for the sample birds the air pressure was dropped. Therefore, it certainly seems as though the sensing of barometric pressure may be the key to how animals are forecasting the weather.

The observation that birds can be affected by air pressure goes back a long way. W.W. Cooke in 1888 suggested that this was how birds migrate away from storms.[163] A similar observation was reported by E.H. Eaton in 1904, who said that pressure changes induced the migration of birds before rain.[164]

Homing pigeons did seem to be able to sense air pressure.[165] Of 12 birds tested, ten of them seemed to be affected by alterations in barometric pressure. The 50% threshold of effect was approximately 10 mm H_2O, which equated to an altitude change of about 10 m. This suggests that such birds are quite sensitive to air pressure and may be able to recognise the onset of a low-pressure climatic region and hence rain. Such a study does give experimental evidence that birds are responding to changes of air pressure, which would explain what is being reported about myths and anecdotes. This was also found by others. In a similar study on ducks (*Anas platyrhynchos*),[166] the authors concluded:

> "The results of this preliminary investigation demonstrate that mallard ducks are capable of discriminating changes as small as 0.4 p.s.i. in atmospheric pressure. Although several sensory systems are used by vertebrates to monitor their environment, the ability to discriminate small changes in atmospheric pressure could play an important role in the daily and seasonal movements of ducks as well as of other species.

In this paper, which includes a photograph of the pressure chamber the authors used, along with an excellent sketch of a duck and what it needed to do in the test – they used an intelligence panel with a light stimulus, two switches and a food tray – the authors suggested that there were multiple reasons why ducks could use their sense of air pressure. This included sensing a forthcoming storm (perhaps six to ten hours ahead of its arrival), but also altitude (within 1000 feet) and flight orientation. They suggest that the latter will allow the duck to increase ground speed without altering its air speed, so making flight more efficient. In the introduction of the paper, they also cite a host of other papers which suggest migration is correlated to air pressure.[167] One of these, published 25th February 1949, tells a story of the flight of red wings (*Turdus musicus* L.; European thrush) which sadly ends with the death of many of them.[168] In response a previous article (H. Landsberg, *Science*, December 24, p708) the author describes the weather conditions and why the birds may have had their migration altered by the barometric conditions.

Lehner and Dennis (1971)[169] cite a list of papers which suggest that birds (and other animals) migrate because of changes in air pressure. This includes six from A.M. Bagg,[170] all from *Audubon Field Notes*. This publication started in July 1947 as a collaboration between the National Audubon Society and the U.S. Fish and Wildlife Service.[171] Others too have quoted Bagg, but not all papers can be taken at face value because they have been cited by others. For example, Curtis (who cites Bagg) does look at air pressure systems and bird migration, but suggests that as a high-pressure ridge passes the birds are responding to "clear skies or winds shifting to southerly or calm".[172] They do not suggest that the birds are sensing barometric pressure *per se*.

Other papers cited by Lehner and Dennis[173] include one by Hicks[174] from Iowa State College on squirrels, and he says that "squirrel activity increases slightly with increases in barometric pressure". However, many other environmental factors were also correlated with animal activity and a direct effect of barometric pressure was not substantiated.

Nisbet and Drury[175] looked at the migration of songbirds and waterbirds, and correlated this to 19 weather variables. When it came to air pressure, they found that "[m]igration density was significantly correlated with high and rising temperature, low and falling pressure", the latter being pertinent here. Richardson[176] looked at the migration of birds into southern Ontario and correlated this to weather patterns, including air pressure. He found that "most migration occurs with a pressure gradient falling from east to west", which suggests that the birds were responding to changing barometric pressure. However, he then goes on to say, in the conclusions, that this was because the pressure area causes warm following winds, and it is this that promotes the bird activity, but not the changes in air pressure.

In 1929, William Rowan published a paper[177] based on him joining a hunting party in Alberta for a "day's duck-shooting on October 30, 1928." When he arrived there were reports that on the 29th there were numerous mallards present and that he "could depend on excellent shooting." However, arriving on the 30th he found that "nowhere were Mallards to be found in sufficient numbers to make even mediocre hunting". This obviously made Rowan wonder what was going on, and why the birds had all disappeared. He was obviously disappointed as in the paper he describes the day as a complete failure, but not for the ducks, obviously. He said that there was a conceived perception that the birds migrate before the onset of winter storms but during what was described as a migration which was "one of the most remarkable ever to have been observed", there was a long spell of very high pressure, and Rowan was wondering if this had an influence. The only other weather aspect that was worth noting was a gently dropping temperature. The conclusion of the paper was that the birds were sensitive to latitude.

Therefore, although not all the evidence given supports the idea that birds can indeed sense barometric pressure directly and have a response to changes, the consensus seems to be that birds can sense barometric pressure and change their behaviour accordingly. Some of the behaviour changes are clearly because the air pressure changes are altering other weather patterns, such as winds, and few of these relatively old papers have direct experimental evidence of barometric pressure effects on birds.

Others have looked at the effects of barometric pressure on other animals, not just birds. Regarding the frog *Cophixalus ornatus*, there appeared to be little effect on air pressure on the croaking behaviour.[178] On the other hand, in another study of five frog species (*Rana sylvatica, Pseudacris crucifer, Bufo americanus, Rana clamitans* and *Rana catesbeiana*) barometric pressure was found to be of increasing importance in the croaking observed, particularly during the latter part of the calling period.[179]

In a study of how fish move into saltmarshes in Victoria, Australia, there was also no influence of barometric pressure.[180] The authors looked at fish abundance and the species variability and neither seemed to be affected. However, sharks (blacktip sharks, *Carcharhinus limbatus*) were thought to move to deep water when a storm was approaching and this was triggered by their sensing a drop of barometric pressure.[181] They returned to shallower water once the storm had passed.

With bats there also did seem to be an effect of pressure. In a study of cave-dwelling bats, it was concluded that the Eastern Pipistrelle bat (*Pipistrellus subflavus*) senses barometric pressure metabolically. It was suggested that this is used to predict the abundance of insects which may be

Blacktip shark (*Carcharhinus limbatus*).
(Sourced from Shutterstock, contributed by Stefan Pircher, but made black and white.)

flying outside the roost.[182] More recently, a study of bat behaviour stated in their conclusions:

> Nightly bat activity was positively related to average nightly temperature and average nightly barometric pressure. In contrast to our expectations, bat activity was not related to changes in barometric pressure prior to or during sample nights.[183]

Again, the authors suggested that this might be related to availability of food and benefits for flight.

At Lanyu (Orchid Island) in Taiwan the sea snakes (*Laticauda* spp.; otherwise referred to as sea kraits) indicate the coming of a storm by disappearing. This was particularly noted before the arrival of a typhoon (Morakot), which caused disruption on the island between 7th and 9th August 2009.[184] It was found that the snakes were not responding to wind speed or precipitation but were sensing the barometric pressure changes prior to the arrival of the cyclone.[185] It was thought that the snakes hide in the crevasses of the volcanic rocks in the area, safely away from the rough water caused by the storm. However, this was an observational study and no direct evidence of the snakes sensing barometric pressure was sought.

To try to understand the correlation of barometric pressure and pain, mice were placed in a chamber and the pressure dropped from 1013 hPa by 40 hPa for 50 minutes. Besides behaviour changes, the expression of

A sea krait (*Laticauda colubrina*).
(Sourced from Shutterstock, contributed by dwi putra stock: This image
has been made black and white.)

the protein *c*-Fos[186] was measured. The authors concluded that neurons in
the superior vestibular nucleus (SuVe) responded to the lowering of pressure.[187] The fact that gene expression is being altered here is significant, as
this suggests a long-term change in the cells, perhaps adapting to the new
environmental condition, i.e. air pressure. This might be a stress response,
but one that would have a positive outcome, enabling the future to be more
secure. This would echo the responses seen in individual animals, where
the reaction of changes can usually be corelated to survival: nest building,
extra feeding, moving to safer ground etc.

In a similar manner, rats were placed in a climate-controlled room where
they had to undergo a forced swim test. This is a recognised method for
assessing the sense of despair in rodents. Atmospheric pressure was lowered
in the range which would be comparable with different weather conditions,
and it was found that such lowing of air pressure increased the depression-
like behaviour in the animals.[188] Such work shows that there are defined
responses to barometric pressure in higher organisms, and therefore the
observations of many species altering their activities in response to changing pressures – perhaps indicative of changing weather – has some rational
basis. If rats and mice are sensitive to barometric pressure, are humans too?
Some have tried to determine this with no success,[189] although the internet
has discussions on this topic,[190] with people reporting joint pain and head-
aches triggered by air pressure changes.

In lower organisms too, barometric pressure seems to be important. In the ectoparasitoid[191] *Mallophora ruficauda*, changes in pressure altered the host-seeking behaviour of the larvae.[192] With effects in a wide range of organisms, it suggests that sensing of barometric pressure was an early evolutionary characteristic, or has developed more than once. Interestingly, early life might have had an Earth with a different atmospheric pressure,[193] and perhaps evolution has had to allow adaptation to different pressures as animals developed. This may then have been adopted for other reasons, such as sensing weather and informing migration patterns, now seen by humans as predictors of future rain or sun.

In a 2013 paper in the journal *PlosOne*[194] it was reported that the mating behaviour of insects was modified in response to air pressure, suggesting that such environmental changes can indeed alter behaviour. The authors looked at three unrelated insect species: curcurbit beetle, *Diabrotica speciosa* (Coleoptera); true armyworm moth, *Pseudaletia unipuncta* (Lepidoptera); potato aphid, *Macrosiphum euphorbiae* (Hemiptera). These animals were observed during periods of naturally occurring air pressure changes, but also when the pressure was manipulated under experimental conditions. Whilst the responses by the different species were not all the same, they all seemed to alter behaviour as the pressure changed. For example, in the armyworm, mating was reduced as the pressure rose, whilst the beetles had less locomotion and were less responsive to female pheromones[195] as pressure dropped. It was suggested that such responses enable the insects to survive better as the weather becomes bad. In a review of this paper, it was suggested that there are hair-like receptors in the cuticle of the insects which are responsible for sensing air pressure, and it was also suggested that this might be widespread in the insects.[196] A similar observation had previously been published in 1993, when observations of a parasitic wasp (*Leptopilina heterotoma*) were made.[197] Even though the egg laying behaviour of the animals was different as the barometric pressure changed, the conclusions made at the time were somewhat undecisive: "We cannot firmly state that the wasps respond only to dropping barometric pressure or to changes in barometric pressure." However, this would explain what was observed, although the authors moot that the insects are sensing change and the egg laying in pre-parasitised prey is a survival response.

Before leaving barometric pressure, at least for now, let us return to the leaf-cutter ants, who we met at the start of Chapter 2. It was reported that these insects increase their rate of foraging if the barometric pressure drops.[198] In a study of *Atta sexdens*, Sujimoto *et al.*[199] noticed that as the air pressure dropped the scout ants were the first to react, leaving the nest more promptly. The foragers then cut more leaf material (up to 1.5 times as much) and then brought back far more leaf cuts, up to twice as much, although it should be noted that the number of foragers did not increase, they just seemed to work harder. This would ensure that the nest was well stocked before the weather turned poor and the trails were no longer passable for collecting leaves. The nest would survive and be able to ride out the storm. However, the ants would need to predict the impending doom far enough in advance to stock the nest.

There seems little doubt that a range of animals can sense barometric pressure and alter their behaviour. This behaviour change is often appropriate to making sure that future survival is maximised. Even effects in mammals are reported, and therefore before the sayings about ants, hedgehogs or cows are dismissed as fables, perhaps there is more of a scientific basis to this than many will credit.

As well as barometric pressure some animals may be sensitive to sound which we do not hear. The idea that storms can produce infrasound has been around a long time.[200] In a 2000 paper by Bedard and Georges,[201] they discuss the emissions of infrasounds, with frequencies from 1 Hz to 20 Hz. They list several natural sources including ocean waves, severe weather, tornadoes, earthquakes, volcanoes, avalanches and even meteors. Georges and Greene[202] pose the question as to whether infrasound is useful for storm warning. They conclude that "emissions show promise", but had a concern about cost-effectiveness of the technologies available to them in 1975. More recently[203] Bedard et al., using some of the data from Bedard and Georges, discuss the use of the Infrasound Network (ISNET) to detect tornadoes. There seems little doubt that some weather systems can give off sound which humans cannot hear. The question is, can animals? And therefore, can they hear a storm which is inaudible to us? As discussed earlier, some animals can. Elephants can hear sound below 20 Hz, much lower than we can.

One of the people who has published a body of work on infrasound and animals is Jonathan T. Hagstrum, of the U.S. Geological Survey. In a 2000 paper[204] he suggested that pigeons were using infrasound (low-frequency sound) cues as part of their navigational systems, and that this was disrupted by the sound output of the supersonic aircraft, Concorde. Concorde, which does not fly anymore following a catastrophic failure and crash in Paris,[205] was known to create a sonic boom, and people in Cornwall would complain about it at the time,[206] as it was not only loud but said to be destructive. Hagstrum thought it was affecting birds, too. He went on to look at how birds may be using infrasonic maps.[207,208]

Hagstrum's work was based on that of others. Hagstrum and Manley[209] say that the infrasound use for bird navigation was discussed by William von Arx (Woods Hole Oceanographic Institution) and Donald R. Griffin in a paper published in 1969.[210] Yodlowski et al.[211] showed that homing pigeons (*Columba livia*) were sensitive to infrasound, an observation confirmed by Kreithen and Quine,[212] who said that pigeons could detect sound down to 0.05 Hz. Hagstrum and Manley reported on the flight of pigeons which had had their cochleae and lagenae destroyed. These birds had a severely disrupted navigation, suggesting the auditory cues are required. Hagstrum went on, in a reinterpretation of work by Gerhart Wagner,[213] to suggest that atmospheric conditions can interrupt the infrasound signals that the birds require.[214] Wagner released his birds above a temperature inversion,[215] and Hagstrum suggested that the birds were then deaf to the sound as it could not penetrate this atmospheric layer. The birds could not navigate properly and failed to return to their lofts.

It is not just in navigation that infrasound can influence animals. Freeman and Hare[216] found that peacocks produce and respond to infrasound during their mating displays. They also point out that others have reported other animals which can produce infrasound, notably elephants (*Loxodonta africana*)[217] and the western capercaillie (*Tetrao urogallus*). Freeman and Hare say that pigeons and guinea fowl (*Numida meleagris*) use infrasound-sensitive neurons which are situated in the midbrain.

It is clear therefore that infrasound needs to be thrown into the mix of environmental conditions that animals can use to sense some changes in weather, such as a storm approaching. Such weather can emit such sound waves, and some animals are capable of hearing them. Along with changes

A western capercaillie (*Tetrao urogallus*).
(Sourced from Shutterstock, contributed by Rudmer Zwerver. This image has been made black and white.)

in barometric pressure, possible changes in temperature and even perhaps the vibrations of the ground, animals may be much more sensitive to changes of conditions which indicate a change in the weather than humans.

The generation of infrasound by natural phenomena is not restricted to the weather, and as Bedard and Georges listed, other environmental events can produce such noises, such as earthquakes, and therefore the perception of infrasound will be revisited in the next chapter, too.

3.7 Chapter summary

What is clear from this roam through the literature is that in many cultures around the world the observations of animals are used as indicators of the weather which is on the way. Some animals are even nicknamed because of their weather-forecasting ability. For example, the yellow-billed cuckoo is often referred to as a rain crow.[218]

This text is not the first compilation of sayings and myths about the weather. "Uncle Offa", as mentioned earlier, had a slot in a radio show and this culminated in a book.[219] The true author, Frederick Hingston, listed many examples of sayings about animals and weather. Many of these, along with others from the discussions earlier, are listed in Table 1. Examples from Offa include, "Bees won't swarm before a storm", "Bees won't leave the hive if it is going to rain", "Spiders' web on long lines foretell a fine day, but if they shorten the threads it will rain." Offa says that this last saying is "totally reliable". The radio show precipitated what appears to be a flurry of comments from listeners and Hingston collected many examples, although the book is mainly focused on relevant days in the year, rather than animals. Even so, it shows how common these sayings are – people clearly engaged with his radio slot.

The range of animals used for weather prediction is large, from ants to cows, but in many cases, there is little or no scientific evidence to support the claims made. Having said that, as pointed out, many of these anecdotes and myths have lasted for generations and clearly there is some continuity of this forecasting. As listed in Table 2, it appears that a range of organisms can sense barometric pressure, and often changes are correlated to altered activities. This is seen from parasites to birds and bats. If an animal is altering its behaviour as a low pressure is sensed, indicative of rain arriving soon, then humans can observe this and get a jump ahead of impending weather changes, as the Tempest Prognosticator showed. Animals can hear sound outside the range perceived by humans, and animals sense changes in temperatures too, so a combination of sensing may well give a prediction of weather changes. It hardly needs to be too accurate too. If a shark swims deep and no storm comes, it has probably not behaved in a life-threatening manner. However, if it does not swim deep, being caught in the shallows may be catastrophic for the shark if a storm does come, so erring on the side of caution may be a good evolutionary characteristic.

What is also clear is that as the climate changes, such weather indicators may be of some value if technological-based weather forecasting is not available, such as in many remote parts of the world. Observations which allow peoples to plant and harvest crops, or batten down against the coming storm, may be critical to the survival of some communities, and if the satellites and computer modelling are unavailable, or simply wrong, the croaking of a frog may be important.

There is no doubt the climate is changing, and unprecedented events are happening. At the time of writing, a glacier just collapsed in Italy, killing several hikers. Some are saying this should have been predicted, whist others are saying it could not have been foreseen. Yet the Italian climber Reinhold Messner reportedly said: "What has happened there is happening every day on every glacier."[220] On a related note, it has been reported that even national borders are not safe from climate change. As the Theodul Glacier retreats, Switzerland and Italy are arguing over where their borders lie. This border is determined by the water runoff and which way it travels, but as the glacier melts, the land underneath has shifted this watershed. Normally, high in the mountains, this may be of little consequence, but with a large ski development at this location there is now an international discussion about how this might be resolved.[221] The weather patterns are changing, and perhaps animal observations can aid in how we understand what is happening.

Changes in weather can have significant consequences, but there are other catastrophic events which are important to forecast too. In the next chapter, the idea of using animals to predict major events in the environment, such as earthquakes, volcanic eruptions and tsunami, will be discussed.

Endnotes

1   Almanac: www.almanac.com/how-insects-predict-weather (Accessed 02/11/21)
2   In March 2022 it was announced that the wreck of Shackleton's ship, *Endurance*, had been found 107 years after it had been crushed and sank in the Weddell Sea. It was amazingly intact, lying at a depth of 3,008 m. The BBC: www.bbc.co.uk/news/science-environment-60662541 (Accessed 22/06/22)
3   Lansing, A. (2000) *Endurance Shackleton's Incredible Voyage*. Weidenfeld & Nicholson, London. ISBN: 9780753809877
4   Clark-Platts, A. (2022) *The Cove* (p. 177). Raven Books, London. ISBN: 9781526604279

5 The Guardian: www.theguardian.com/business/2019/jul/25/why-the-heatwave-is-disrupting-the-uk-railways (Accessed 14/03/22)

6 The Independent: www.independent.co.uk/extras/lifestyle/british-people-time-spent-talking-weather-conversation-topic-heatwave-a8496166.html (Accessed 02/11/21)

7 Bristol Airport: www.bristolairport.co.uk/about-us/news-and-media/news-and-media-centre/2018/8/brits-and-weather-research (Accessed 13/09/22)

8 The broadcast can be seen on YouTube at: www.youtube.com/watch?v=NnxjZ-aFkjs (Accessed 04/11/21)

9 The BBC: www.bbc.co.uk/news/av/uk-england-41616367 (Accessed 04/11/12)

10 The BBC: www.bbc.co.uk/news/magazine-32483678 (Accessed 02/11/21)

11 Darwin, C. (1859) *On the Origin of Species by Means of Natural Selection, or the Preservation of Favoured Races in the Struggle for Life.* John Murray, London.

12 The Metrological Office: www.metoffice.gov.uk/ (Accessed 02/11/21)

13 There are now two major motorway bridges next to each other. It is the old one I am mainly referring to.

14 National Highways: https://nationalhighways.co.uk/travel-updates/the-severn-bridges/ (Accessed 22/06/22)

15 Hart, C. (1969) Mediaeval kites and windsocks. *The Aeronautical Journal*, 73(708), 1019–1026.

16 Robinson, B.R. and Johnson, D. (2009) For emergency medical service helicopter pilots, all wind is local. *Air Medical Journal*, 28(5), 256.

17 Mulinazzi, T.E. and Schrock, S. Ph.D., P.E. (2010) Mitigating wind induced truck crashes. *Final Reports & Technical Briefs from Mid-America Transportation Center*, 98. https://digitalcommons.unl.edu/matcreports/98 (Accessed 02/11/21)

18 Guse, B., Anwar, F., Merz, B., Tarasova, L., Merz, R., Bárdossy, A. and Vorogushyn, S. (2021) Event indicator analysis using depth functions to explain the occurrence of large floods in Germany. In *EGU General Assembly Conference Abstracts* (pp. EGU21–14692).

19 Euronews: www.euronews.com/2021/09/23/i-know-lots-of-people-that-don-t-want-to-return-how-flooding-is-threatening-to-empty-germa (Accessed 02/11/21)

20 CNN: https://edition.cnn.com/2021/10/12/china/flooding-china-shanxi-ntl-hnk/index.html (Accessed 02/11/21)

21 ABC News: https://abcnews.go.com/International/wireStory/heavy-rains-south ern-india-kill-14-people-flood-81104852 (Accessed 12/11/21)

22 Jiang, W., Wang, L., Zhang, M., Yao, R., Chen, X., Gui, X., Sun, J. and Cao, Q. (2021) Analysis of drought events and their impacts on vegetation productivity based on the integrated surface drought index in the Hanjiang River Basin, China. *Atmospheric Research*, 254, 105536.

23 Mohsin, M. and Pilz, J. (2021) Stochastic model for drought analysis of the Colorado River Basin. *Stochastic Environmental Research and Risk Assessment*, 1–12.

24 IPCC Sixth Assessment Report: www.ipcc.ch/report/ar6/wg1/ (Accessed 23/06/22)

25 Wakeman, R. (2021) American utopia and climate change. *Amerikastudien*, 66, 279–282.

26 Britannica: www.britannica.com/place/Maldives (Accessed 22/12/21)

27 Britannica: www.britannica.com/place/Marshall-Islands (Accessed 22/12/21)

28 Bedritskii, A.I., Vil'fand, R.M., Kiktev, D.B. and Rivin, G.S. (2017) For numerical weather prediction. *Russian Meteorology and Hydrology*, 42(7), 425–434.

29 Spectrum News 1: https://spectrumnews1.com/wi/green-bay/weather/2020/10/08/wisconsin-weather-blog-meteorologist-wrong-rudd (Accessed 23/06/22)

30 CBC Kids: www.cbc.ca/kidscbc2/the-feed/8-animals-that-give-the-weather-report (Accessed 23/06/22)

31 The jet stream consists of high-level air currents which can dictate the movement of weather patterns at lower altitudes. Also see: Woollings, T. and Blackburn, M. (2012) The North Atlantic jet stream under climate change and its relation to the NAO and EA patterns. *Journal of Climate*, 25(3), 886–902.

32 The Meteorological Office: www.metoffice.gov.uk/weather/warnings-and-advice/uk-storm-centre/index (Accessed 22/12/21)

33 Knowles Middleton, W.E. (1944) A brief history of the barometer. *Journal of the Royal Astronomical Society of Canada*, 38, 41. Available through Harvard: http://adsabs.harvard.edu/pdf/1944jrasc..38 . . . 41k (Accessed 23/06/22)

34 Langford, N.J. and Ferner, R.E. (1999) Toxicity of mercury. *Journal of Human Hypertension*, 13(10), 651–656.

35 United States Lighthouse Society: https://uslhs.org/lens-rotation (Accessed 23/06/22)

36 It is thought that mercury poisoning from the vapours caused many lighthouse keepers to have mental health issues. Originally this was put down to loneliness and confinement, especially on rock lighthouse far out to sea, but more recently it was suggested that mercury could have been the root cause. Most lighthouses now are automated and have no keepers.

37 If anyone were mad enough to try to repeat this, the end of the glass tube must be placed on a solid surface to stop the momentum of the falling mercury from going right through the end of the tube.

38 Mercury clean-up kits are now sold, and are often based on mercury binding to sulphur power, itself not very nice to handle. For example: Ice Cleaning: www.ice cleaning.co.uk/mercury-cleanup-and-disposal?campaign=11411550620&content=473935494655&keyword=mercury%20spill%20cleanup&infinity=ict2~net~gaw~ar~473935494655~kw~mercury%20spill%20cleanup~mt~p~cmp~Mercury%20Spill~ag~Mercury%20Spillage&gclid=Cj0KCQiA2sqOBhCGARIsAPuPK0hSre1s2uPOcSi0NTndLm72kCntb59kLyDkPbb-21HlLr8jIFBaX3EaAsHMEALw_wcB (Accessed 23/06/22)

39 Mac Tutor: https://mathshistory.st-andrews.ac.uk/Biographies/Hooke/ (Accessed 23/06/22)

40 Bryden, D.J. (1975) Sir Samuel Morland's account of the balance barometer, 1678. *Annals of Science*, 32(4), 359–368.

41 Vauxhall History: https://vauxhallhistory.org/sir-samuel-morland/ (Accessed 23/06/22)

42 NNDB: www.nndb.com/people/573/000097282/ (Accessed 23/06/22)

43 The Royal Society: https://royalsocietypublishing.org/doi/10.1098/rstl.1698.0002 (Accessed 02/11/21)

44 The Royal Society: https://royalsocietypublishing.org/doi/10.1098/rstl.1698.0027 (Accessed 02/11/21)

45 National Geographic: www.nationalgeographic.org/encyclopedia/barometer/ (Accessed 02/11/21)

46 Otherwise referred to as udometer (a rather archaic term), pluviometer (*pluvia* is Latin for rain), ombrometer (from the word *umbra* in Latin), and hyetometer (derived from Greek for rain).

47 Valley Forge Cupolas: www.valleyforgecupolas.com/blog/the-history-of-weath ervanes (Accessed 02/11/21)

48 ThoughtCo: www.thoughtco.com/history-of-the-anemometer-1991222 (Accessed 02/11/21)

49 For example, from Gill Instruments: https://skyview-systems.co.uk/products/wo65-windobserver-wind-speed-sensor (Accessed 03/08/22)

50 Atlantic Scale Company, Inc.: https://atlanticscale.com/measuring-humidity-hygrometer-calibration/#:~:text=The%20first%20crude%20hygrometer%20was,meteorological%20devices%20including%20the%20hygrometer (Accessed 08/08/22)

51 Museo Galileo: https://catalogue.museogalileo.it/biography/FrancescoFolli.html (Accessed 08/08/22)

52 Museo Galileo: https://catalogue.museogalileo.it/biography/HoraceBenedictDeSaussure.html (Accessed 08/08/22)

53 Accuweather: www.accuweather.com/en/gb/bristol/bs1-6/hair-day-weather/327328 (Accessed 08/08/22). This URL was based on the address of UWE, Bristol.

54 Britannia: www.britannica.com/science/atmospheric-science (Accessed 08/08/22)

55 Museo Galileo: https://catalogue.museogalileo.it/multimedia/HygrometerBis.html (Accessed 08/08/22)

56 Bresser Group of Companies: www.bresser.de/en/Weather-Time/BRESSER-WIFI-ClearView-Weather-Center-with-7-in-1-Sensor.html (Accessed 03/08/22)

57 AWEKAS: www.awekas.at/wp/?lang=en (Accessed 03/08/22)

58 WeatherCloud: https://weathercloud.net/en (Accessed 03/08/22)

59 Weather Underground: www.wunderground.com/about/our-company (Accessed 03/08/22)

60 Sabeena, J. and Reddy, P.V.S. (2017) A review of weather forecasting schemes. *i-manager's Journal on Pattern Recognition*, 4(2), 27.

61 Pelton, J.N. (2019) The new capabilities of weather satellites. In *Space 2.0* (pp. 59–69). Springer, Cham.

62 Goldbaum, E. (2020) Seven small satellites reimagining how we see weather. *Weatherwise*, 73(4), 24–31.

63 UN Climate Change Conference UK 2021: https://ukcop26.org/ (Accessed 02/11/21)

64 Britannia: www.britannica.com/science/weather-forecasting/History-of-weather-forecasting (Accessed 02/11/21)

65 Ironically, not everyone wished for the dangerous reefs for be marked with warnings, such as bells and lights. When ships were wrecked, locals would collect the spoils, including the cargoes. It was a good source of local income for these communities, and "wreckers" even put up false lights to lure ships to their doom. Therefore, there was some objection to the building of lighthouses. Cornwall has a long history of "wrecking": Pearce, C.J. (2010) *Cornish Wrecking, 1700–1860: Reality and Popular Myth*. Boydell Press, Woodbridge, Suffolk. ISBN: 978-1843835554

66 ICE: www.ice.org.uk/what-is-civil-engineering/civil-engineer-profiles/john-rennie (Accessed 26/02/22)

67 Approximately two thirds of the original Smeaton lighthouse was brought back to Plymouth, in what was a crowd-funding scheme at the time. It is now a visitors' attraction at The Hoe, Plymouth: www.visitplymouth.co.uk/things-to-do/smeatons-tower-p258003 (Accessed 26/02/22)

68 ICE: www.ice.org.uk/what-is-civil-engineering/civil-engineer-profiles/john-smeaton (Accessed 26/02/22)

69   The Ministry of History: www.theministryofhistory.co.uk/short-histories-blog/
     henrywinstanleyandthelighthouseateddystone (Accessed 23/06/22)

70   This painting of the Bell Rock by Turner is at the National Galleries Scotland and
     can be viewed here: www.nationalgalleries.org/art-and-artists/19251/bell-rock-
     lighthouse (Accessed 31/08/22)

71   Bathhurst, B. (2020) *The Lighthouse Stevensons*. HarperCollins Publishers, New
     York. ISBN: 978-0007204434

72   Bathhurst (2020)

73   https://penelope.uchicago.edu/Thayer/E/Roman/Texts/Theophrastus/De_sig
     nis*.html (Accessed 02/11/21)

74   From: Theophrastus of Eresus: On weather signs – Bryn Mawr classical review:
     https://bmcr.brynmawr.edu/2007/2007.09.40/ (Accessed 02/11/21)

75   From https://bmcr.brynmawr.edu/2007/2007.09.40/ (Accessed 02/11/21)

76   Ewer, D.W. (1952) Animal weather prophets. *Weather*, 7, 16–19.

77   Manyanhaire, I.O. (2015) Integrating indigenous knowledge systems into cli-
     mate change interpretation: Perspectives relevant to Zimbabwe. *Greener Journal of
     Educational Research*, 5, 27–36.

78   Enock, C.M. (2013) Indigenous knowledge systems and modern weather forecast-
     ing: Exploring the linkages. *Journal of Agriculture and Sustainability*, 2(2).

79   Enock (2013).

80   Risiro, J., Mashoko, D., Tshuma, Doreen, T. and Rurinda, E. (2012) Weather fore-
     casting and indigenous knowledge systems in Chimanimani District of Manicaland,
     Zimbabwe. *Journal of Emerging Trends in Educational Research and Policy Studies*, 3(4),
     561–566.

81   Radeny, M., Desalegn, A., Mubiru, D., Kyazze, F., Mahoo, H., Recha, J., Kimeli,
     P. and Solomon, D. (2019) Indigenous knowledge for seasonal weather and climate
     forecasting across East Africa. *Climatic Change*, 156(4), 509–526.

82   Some regions of the world have two rainy seasons. There is the long, or big, rains
     from April to May, and then the short, or small, rains in October to December.

83   Janzen, F.J. (1994) Climate change and temperature-dependent sex determination
     in reptiles. *Proceedings of the National Academy of Sciences*, 91(16), 7487–7490.

84   Radeny *et al.* (2019).

85   Okonya, J.S. and Kroschel, J. (2013) Indigenous knowledge of seasonal weather
     forecasting: A case study in six regions of Uganda. *Agricultural Sciences*, 2013.

86   Aotearoa is the Māori name for New Zealand.

87   NIWA: https://niwa.co.nz/sites/niwa.co.nz/files/Traditional-Maori-Weather-and-
     Climate-Forecasting-poster.pdf (Accessed 02/11/21)

88   Pelster, B. (2004) pH regulation and swimbladder function in fish. *Respiratory
     Physiology & Neurobiology*, 144(2–3), 179–190.

89   Pets on Mom: https://animals.mom.com/why-is-my-goldfish-eating-gravel-125
     64419.html (Accessed 22/12/21)

90   Prober, S.M., O'Connor, M.H. and Walsh, F.J. (2011) Australian Aboriginal peo-
     ples' seasonal knowledge: A potential basis for shared understanding in environ-
     mental management. *Ecology and Society*, 16(2).

91   Independent: www.independent.co.uk/sport/olympics/how-beijing-used-rockets-
     to-keep-opening-ceremony-dry-890294.html (Accessed 22/12/21)

92   Tessendorf, S.A., Bruintjes, R.T., Weeks, C., Wilson, J.W., Knight, C.A., Roberts,
     R.D., Peter, J.R., Collis, S., Buseck, P.R., Freney, E. and Dixon, M. (2012) The

Queensland cloud seeding research program. *Bulletin of the American Meteorological Society*, 93(1), 75–90.

93 Prober *et al.* (2011).

94 Lantz, T.C. and Turner, N.J. (2003) Traditional phenological knowledge of Aboriginal peoples in British Columbia. *Journal of Ethnobiology*, 23(2), 263–286.

95 Lawrence, A. (2009) The first cuckoo in winter: Phenology, recoding, credibility and meaning in Britain. *Global Environmental Change*, 19, 173–179.

96 Science Focus: www.sciencefocus.com/nature/is-it-true-that-cows-lie-down-when-its-about-to-rain/ (Accessed 02/11/21)

97 Ewer (1952).

98 Haskell, M.J., Masłowska, K., Bell, D.J., Roberts, D.J. and Langford, F.M. (2013) The effect of a view to the surroundings and microclimate variables on use of a loafing area in housed dairy cattle. *Applied Animal Behaviour Science*, 147(1–2), 28–33.

99 www.thetimes.co.uk/article/cows-lying-down-before-rain-it-stands-up-qwm26 tcs7wg (Accessed 02/11/21)

100 Radeny *et al.* (2019).

101 Wallisch, K. (1999) *Animal Behavior as a Weather Predictor* (Doctoral dissertation, Oklahoma State University). Although Google Scholar cites this as a Doctoral thesis, it was in fact submitted for a Master of Science.

102 Ewer (1952).

103 Okonya and Kroschel (2013).

104 Seagull: there is no such species as a seagull and it is generic term used commonly for gulls which are seen at the coast or out to sea. The term would encompass black-headed gulls, herring gulls etc. For more information see the Royal Society for the Protection of Birds (RSPB): www.rspb.org.uk/birds-and-wildlife/wildlife-guides/bird-a-z/gulls-and-terns/ (Accessed 16/06/22)

105 Sloely: https://sloely.com/seagulls-can-tell-the-weather/ (Accessed 24/06/22)

106 The original aircraft used by the display team the Red Arrows was the Gnat (The Folland Gnat). It was replaced by the Hawk (BAE Systems) in 1980. Both planes were used also for RAF training. Royal Air Force (UK): www.raf.mod.uk/display-teams/red-arrows/the-team/ (Accessed 05/07/22)

107 Ewer (1952).

108 Moffett, *Theater of Insects* (1658; reprint facsimile ed., Da Capo Press, New York, 1967).

109 Bourque, M. (1999) "There is nothing more divine than these, except man": Thomas Moffett and insect sociality. *Quidditas*, 20(1), 9.

110 Origins if modernity: Natural history: https://web.archive.org/web/20060825 154523/www.library.usyd.edu.au/libraries/rare/modernity/moffett.html (Accessed 03/11/21)

111 Lapham's Quarterly: www.laphamsquarterly.org/night/charts-graphs/goodnight-room (Accessed 03/11/21)

112 Sir Humphry Davy was an extremely famous Cornish scientist (from Penzance, where there is still a school named after him), who invented the "Davy lamp" which saved thousands of miners' lives. For a good account of his work see: Holmes, R. (2008) *The Age of Wonder: How the Romantic Generation Discovered the Beauty and Terror of Science*. Harper Press, Toronto, Canada. ISBN: 9780007149537

113 Royal Meteorological Society: www.rmets.org/metmatters/behind-folklore-swallows-flying-high-do-high-flying-swallows-mean-dry-weather (Accessed 02/11/21)

114 Royal Meteorological Society: www.rmets.org/metmatters/behind-folklore-swal
lows-flying-high-do-high-flying-swallows-mean-dry-weather (Accessed 02/11/21)

115 Ewer (1952).

116 Present day Istanbul – the name officially changed in 1930. History: www.history.
com/topics/middle-east/constantinople#section_9 (Accessed 03/11/21)

117 Oxford Bibliographies: www.oxfordbibliographies.com/view/document/obo-
9780195396584/obo-9780195396584-0225.xml (Accessed 03/11/21)

118 The Medieval Bestiary: http://bestiary.ca/prisources/psdetail1611.htm (Accessed
03/11/21)

119 Gilbert White's House and Gardens: www.gilbertwhiteshouse.org.uk/gilbert-
white/ (Accessed 03/11/21)

120 Berenbaum, M. (2008) Entomological bandwidth. *American Entomologist*, 54(4), 196–197.

121 Sloane, E. (2005) *Eric Sloane's Weather Book*. Courier Corporation, North
Chelmsford, MA [page 4, which is publicly available].

122 Almanac: www.almanac.com/woolly-bear-caterpillars-and-weather-prediction
(Accessed 02/11/21)

123 Almanac: www.almanac.com/woolly-bear-caterpillars-and-weather-prediction
(Accessed 02/11/21)

124 Indystar/USA Today: https://eu.indystar.com/story/news/environment/2021/09/
27/indiana-weather-woolly-bear-caterpillar-predictions-weather-folklore-other-
facts/5816505001/ (Accessed 02/11/21)

125 Sadewasser, J. (1976) *The Reliability of Selected Weather Beliefs* (Masters theses &
Specialist Projects). Paper 2817. Western Kentucky University: https://digitalcom
mons.wku.edu/theses/2817 (Accessed 02/11/21)

126 Layne Jr, J.R., Edgar, C.L. and Medwith, R.E. (1999) Cold hardiness of the
woolly bear caterpillar (*Pyrrharctia isabella* Lepidoptera: Arctiidae). *American Midland
Naturalist*, 293–304.

127 Yi, S.X. and Lee Jr, R.E. (2016) Cold-hardening during long-term acclimation in
a freeze-tolerant woolly bear caterpillar. *Pyrrharctia isabella. Journal of Experimental
Biology*, 219(1), 17–25.

128 Duman, J.G., Bennett, V., Sformo, T., Hochstrasser, R. and Barnes, B.M. (2004)
Antifreeze proteins in Alaskan insects and spiders. *Journal of Insect Physiology*, 50(4),
259–266.

129 Davies, P.L., Baardsnes, J., Kuiper, M.J. and Walker, V.K. (2002) Structure and
function of antifreeze proteins. *Philosophical Transactions of the Royal Society of
London. Series B: Biological Sciences*, 357(1423), 927–935.

130 DeVries, A.L. (2020) Fish antifreeze proteins. In *Antifreeze Proteins Volume 1*
(pp. 85–129). Springer, Cham.

131 Greene, A.E., Athreya, B., Lehr, H.B. and Coriell, L.L. (1967) Viability of cell
cultures following extended preservation in liquid nitrogen. *Proceedings of the Society
for Experimental Biology and Medicine*, 124(4), 1302–1307.

132 History: www.history.co.uk/shows/mountain-men/articles/the-day-sir-ranulph-fien
nes-lost-his-fingers (Accessed 22/12/21)

133 Sadewasser (1976).

134 Wallisch (1999). Although Google Scholar cites this as a doctoral thesis, it was in
fact submitted for a Master of Science.

135 Garriott, E.B. (1903) Weather folk-lore and local weather signs. (Reprint of
Government Printing Office United States Department of Agriculture Weather
Bureau Bulletin; Grand River Books, Detroit, 1971): as cited by Wallisch (1999).

136 Ewer (1952).

137 The Jenner Institute: www.jenner.ac.uk/about/edward-jenner (Accessed 15/12/21)

138 COVID-19 is caused by a virus known as SARS-CoV-2, which is related to the virus which caused SARS in China in 2002 and 2004. For a good book exploring the origins of the SARS-CoV-2 virus see: Chan, A. and Ridley, M. (2021) *Viral: The Search for the Origin of Covid-19*. 4th Estate, London. ISBN: 9780008487492

139 Sachdeva, A. and Saha, A. (2021) COVID-19 vaccine: A way out of crisis. Available at IntechOpen: www.intechopen.com/chapters/77533 (Accessed 15/12/21)

140 Bardell, D. (1996) Nestling cuckoos to vaccination: A commemoration of Edward Jenner. *BioScience*, 46(11), 866–871.

141 Ewer (1952).

142 Atlas Obscura: www.atlasobscura.com/articles/leeches-predict-weather-tempest-prognosticator (Accessed 14/12/21)

143 Atlas Obscura: www.atlasobscura.com/articles/leeches-predict-weather-tempest-prognosticator (Accessed 24/03/22)

144 The Guardian: www.theguardian.com/news/2015/apr/19/weatherwatch-forecasting-tempest-prognosticator-storm-leech (Accessed 15/12/21) [this contains a photograph of a Tempest Prognosticator].

145 *Great Exhibition of the Works of Industry of All Nations*, which was at the Crystal Palace in Hyde Park, London.

146 Whitby Museum: https://whitbymuseum.org.uk/ (Accessed 22/12/21)

147 Wellcome Collection: https://wellcomecollection.org/works/pq5z9qfv/items?canvas=3 (Accessed 04/08/22)

148 Sattler, H.R. (1978) *Nature's Weather Forecasters*. Thomas Nelson, New York.

149 Ewer (1952).

150 De la Garza, E. (1972) Newsletter-1972–08–17. Available at University of Texas Rio Grande Valley: https://scholarworks.utrgv.edu/cgi/viewcontent.cgi?article=1107&context=kikadelagarzanews (Accessed 22/12/21)

151 BIOS/Bermuda Institute of Ocean Sciences: www.bios.edu/currents/the-science-of-shark-oil-barometers (Accessed 22/12/21)

152 Atlas Obscura: www.atlasobscura.com/articles/predicting-the-weather-with-shark-oil (Accessed 22/12/21)

153 Google Books: https://books.google.co.uk/books?id=j-J4Rz5zlxYC&pg=PA131&dq=shark+oil+barometer&hl=en&sa=X&redir_esc=y#v=onepage&q=shark%20oil%20barometer&f=false (Accessed 22/12/21)

154 SOFAR Bermuda: http://sofarbda.org/thatcher-adams.html (Accessed 22/12/21)

155 Clark-Platts, A. (2022) *The Cove* (p. 177). Raven Books, London. ISBN: 9781526604279

156 Telegraph: www.telegraph.co.uk/environment/2022/08/17/what-smell-rain-called-answer-petrichor-intrigues-scientists/ (Accessed 30/08/22)

157 Gerber, N.N. and Lechevalier, H.A. (1965) Geosmin, an earthy-smelling substance isolated from actinomycetes. *Applied Microbiology*, 13, 935–938.

158 Ackerman, J. (2016) *The Genius of Birds*. Corsair, London. ISBN: 9781472114365

159 Clark-Platts (2022).

160 Sciencing: https://sciencing.com/seagull-behaviors-earthquakes-changes-weather-23098.html (Accessed 02/11/21)

161 Radhouani, H., Igrejas, G., Pinto, L., Gonçalves, A., Coelho, C., Rodrigues, J. and Poeta, P. (2011) Molecular characterization of antibiotic resistance in enterococci

recovered from seagulls (*Larus cachinnans*) representing an environmental health problem. *Journal of Environmental Monitoring*, 13(8), 2227–2233.

162 Breuner, C.W., Sprague, R.S., Patterson, S.H. and Woods, H.A. (2013) Environment, behavior and physiology: Do birds use barometric pressure to predict storms? *Journal of Experimental Biology*, 216(11), 1982–1990.

163 Cooke, W.W. (1888) Report on bird migration in the Mississippi Valley in the years 1884 and 1885. *U.S.D.A. Division Economic Ornithology Bulletin*, No. 2, 16–25.

164 Eaton, E.H. (1904) Spring bird migrations of 1903. *Auk*, 21(3), 341–345.

165 Kreithen, M.L. and Keeton, W.T. (1974) Detection of changes in atmospheric pressure by the homing pigeon. *Columba livia. Journal of Comparative Physiology*, 89(1), 73–82.

166 Lehner, P.N. and Dennis, D.S. (1971) Preliminary research on the ability of ducks to discriminate atmospheric pressure changes. *Annals of the New York Academy of Sciences*, 188(1), 98–109.

167 Listed at the end of this chapter.

168 Suffern, C. (1949) Pressure patterns in bird migration. *Science*, 109, 209.

169 Lehner and Dennis (1971), 98–109.

170 Bagg, A.M. (1960) A summary of the spring migration season. *Audubon Field Notes*, 14, 360–364. Bagg, A.M. (1963) Spring migration, 1963. *Audubon Field Notes*, 17, 380–383. Bagg, A.M. (1964) A diversity of observations for a variety of ornithological tastes (spring migration). *Audobon Field Notes*, 18, 420425. Bagg, A.M. (1965) Spring migrants; the few and the many. *Audubon Field Notes*, 19, 438–446. Bagg, A.M. and Bagg, T. (1962) Spring migration. *Audubon Field Notes*, 16, 382–386 Bagg, A.M. and Baird, J. (1961) A summary of the 1961 spring migration. *Audubon Field Notes*, 15, 380–389.

171 University of California Press: https://online.ucpress.edu/abt/article-abstract/11/1/24/3793/Audubon-Field-Notes?redirectedFrom=fulltext (Accessed 08/08/22)

172 Curtis, S.G. (1969) Spring migration and weather at Madison, Wisconsin. *The Wilson Bulletin*, 81, 235–245.

173 Many of the papers cited by Lehner and Dennis could not be accessed for this discussion. This was because they are relatively old and not digitised or made accessible. These include: Smith, F. (1918) Bird migration and the weather. *Illinois Audubon Society Bulletin*, 15–17; Robbins, C.S. (1949) Weather and bird migration. *Wood Thrush*, 4, 130–144; Wagner, H.O. (1957) The technical basis of experimental research on bird migration. *Ibis*, 99, 191–195; Nisbet, I.C.T. (1959) Migration at Ithaca in spring, 1958. *Kingbird*, 8, 102–104; Newman, R.J. and Lowery, G.H. Jr. (1959) A summary of the 1959 spring migration and its geographic background. *Audubon Field Notes*, 13, 346–252; Johnson, J.W., Pack, A.B. and Jonkel, G. (1961) A spring migration of waterfowl in briefly favorable weather. *South Dakota Bird Notes*, 13, 64–70.

174 Hicks, E.A. (1949) Ecological factors affecting the activity of the western fox squirrel, *Sciurus niger rufiventer* (Geoffroy). *Ecological Monographs*, 19, 287–302.

175 Nisbet, I.C. and Drury Jr, W.H. (1968) Short-term effects of weather on bird migration: A field study using multivariate statistics. *Animal Behaviour*, 16, 496–530.

176 Richardson, W.J. (1966) Weather and late spring migration of birds into southern Ontario. *The Wilson Bulletin*, 78, 400–414.

177 Rowan, W. (1929) Migration in relation to barometric and temperature changes. Bulletin of the Northeastern Bird-Banding Association. *Bird-Banding*, 5, 85–92.

178 Brooke, P.N., Alford, R.A. and Schwarzkopf, L. (2000) Environmental and social factors influence chorusing behaviour in a tropical frog: Examining various temporal and spatial scales. *Behavioral Ecology and Sociobiology*, 49(1), 79–87.

179 Oseen, K.L. and Wassersug, R.J. (2002) Environmental factors influencing calling in sympatric anurans. *Oecologia*, 133(4), 616–625.

180 Crinall, S.M. and Hindell, J.S. (2004) Assessing the use of saltmarsh flats by fish in a temperate Australian embayment. *Estuaries*, 27(4), 728–739.

181 Heupel, M.R., Simpfendorfer, C.A. and Hueter, R.E. (2003) Running before the storm: Blacktip sharks respond to falling barometric pressure associated with Tropical Storm Gabrielle. *Journal of Fish Biology*, 63(5), 1357–1363.

182 Paige, K.N. (1995) Bats and barometric pressure: Conserving limited energy and tracking insects from the roost. *Functional Ecology*, 463–467.

183 Bender, M.J. and Hartman, G.D. (2015) Bat activity increases with barometric pressure and temperature during autumn in central Georgia. *Southeastern Naturalist*, 14(2), 231–242.

184 Liu, Y.L., Lillywhite, H.B. and Tu, M.C. (2010) Sea snakes anticipate tropical cyclone. *Marine Biology*, 157(11), 2369–2373.

185 The word cyclone is a little confusing. A cyclone is a tropical storm in the South Pacific and Indian Ocean. In the Northwest Pacific Ocean, they would be referred to as typhoons, whilst hurricanes occur in the North Atlantic Ocean and Northeast Pacific (The BBC: www.bbc.co.uk/newsround/24879162 (Accessed 04/11/21). However, people talk about cyclonic regions (cyclones: low pressure) and anticyclones (high pressure).

186 *c*-Fos is a transcription factor so alters the expression of genes in cells: Gallo, F.T., Katche, C., Morici, J.F., Medina, J.H. and Weisstaub, N.V. (2018) Immediate early genes, memory and psychiatric disorders: Focus on *c*-Fos, Egr1 and Arc. *Frontiers in Behavioral Neuroscience*, 12, 79.

187 Sato, J., Inagaki, H., Kusui, M., Yokosuka, M. and Ushida, T. (2019) Lowering barometric pressure induces neuronal activation in the superior vestibular nucleus in mice. *PLoS One*, 14(1), e0211297.

188 Mizoguchi, H., Fukaya, K., Mori, R., Itoh, M., Funakubo, M. and Sato, J. (2011) Lowering barometric pressure aggravates depression-like behavior in rats. *Behavioural Brain Research*, 218, 190–193.

189 Staut, A.J. (2001) The effects of barometric pressure on elementary school students' behavior. Available at University of Wisconsin-Stout: https://minds.wisconsin.edu/bitstream/handle/1793/40181/2001stauta.pdf?sequence=1 (Accessed 22/12/21). A repeat of this study by another author also concluded the same. University of Wisconsin-Stout: https://minds.wisconsin.edu/bitstream/handle/1793/42237/2007blaskowskin.pdf?sequence=1 (Accessed 22/12/21)

190 MedicineNet: www.medicinenet.com/how_does_barometric_pressure_affect_humans/article.htm (Accessed 22/12/21)

191 An ectoparasitoid: a parasite which lives externally on its host.

192 Crespo, J.E. and Castelo, M.K. (2012) Barometric pressure influences host-orientation behavior in the larva of a dipteran ectoparasitoid. *Journal of Insect Physiology*, 58(12), 1562–1567.

193 Astrobiology Web: http://astrobiology.com/2020/03/earths-ancient-barometric-pressure.html (Accessed 02/11/21)

194 Pellegrino, A.C., Peñaflor, M.F.G.V., Nardi, C., Bezner-Kerr, W., Guglielmo, C.G., Bento, J.M.S. and McNeil, J.N. (2013) Weather forecasting by insects:

Modified sexual behaviour in response to atmospheric pressure changes. *PloS One*, 8(10), e75004.

195 A pheromone is akin to a hormone, but is released by an individual of a species to be sensed by a different individual of that species, whereas a hormone stays inside an individual body.

196 Nature: www.nature.com/articles/nature.2013.13874 (Accessed 16/11/21)

197 Roitberg, B.D., Sircom, J., Roitberg, C.A., van Alphen, J.J. and Mangel, M., 1993. Life expectancy and reproduction. *Nature*, 364(6433), 108.

198 Phys.Org: https://phys.org/news/2019-12-leafcutter-ants-stormy-weather.html (Accessed 21/12/21)

199 Sujimoto, F.R., Costa, C.M., Zitelli, C.H. and Bento, J.M.S. (2020) Foraging activity of leaf-cutter ants is affected by barometric pressure. *Ethology*, 126(3), 290–296. The paper also includes a nice schematic of the experimental setup (Figure 1).

200 Georges, T.M. (1973) Infrasound from convective storms: Examining the evidence. *Reviews of Geophysics*, 11(3), 571–594.

201 Bedard, A. and Georges, T. (2000) Atmospheric infrasound. *Acoustics Australia*, 28(2), 47–52.

202 Georges, T.M. and Greene, G.E. (1975) Infrasound from convective storms. Part IV. Is it useful for storm warning? *Journal of Applied Meteorology and Climatology*, 14(7), 1303–1316.

203 Bedard Jr, A.J., Bartram, B.W., Keane, A.N., Welsh, D.C. and Nishiyama, R.T. (2004) The infrasound Network (ISNET): Background, design details, and display capability as an 88D adjunct tornado detection tool. In *22nd Conf. On Severe Local Storms*: www.researchgate.net/profile/Randll-Nishiyama/publication/237259557_The_infra-sound_Network_ISNET_Background_design_details_and_display_capability_as_an_88D_adjunct_tornado_detection_tool/links/53f8013e0cf2823e5bdbda0b/The-infrasound-Network-ISNET-Background-design-details-and-display-capability-as-an-88D-adjunct-tornado-detection-tool.pdf (Accessed 05/07/22)

204 Hagstrum, J.T. (2000) Infrasound and the avian navigational map. *Journal of Experimental Biology*, 203, 1103–1111.

205 Britannia: www.britannica.com/topic/Air-France-flight-4590 (Accessed 09/08/22). This URL also has a dramatic picture of the Concorde failing.

206 Hansard: https://api.parliament.uk/historic-hansard/commons/1978/mar/23/concorde-sonic-boom (Accessed 09/08/22)

207 Hagstrum, J.T. (2013) Atmospheric propagation modeling indicates homing pigeons use loft-specific infrasonic 'map' cues. *Journal of Experimental Biology*, 216, 687–699.

208 Hagstrum, J.T. (2013) An infrasound-based avian navigational 'map'. In *Proceedings of Meetings on Acoustics ICA2013* (Vol. 19, No. 1, p. 030088). Acoustical Society of America, Melville, NY, June.

209 Hagstrum, J.T. and Manley, G.A. (2015) Releases of surgically deafened homing pigeons indicate that aural cues play a significant role in their navigational system. *Journal of Comparative Physiology A*, 201, 983–1001.

210 Griffin, D.R. (1969) The physiology and geophysics of bird navigation. *Quarterly Reviews in Biology*, 44, 255–276.

211 Yodlowski, M.L., Kreithen, M.L. and Keeton, W.T. (1977) Detection of atmospheric infrasound by homing pigeons. *Nature*, 265, 725–726.

212 Kreithen, M.L. and Quine, D.B. (1979) Infrasound detection by the homing pigeon: A behavioral audiogram. *Journal of Comparative Physiology*, 129, 1–4.

213 Wagner, G. (1978) Homing pigeons' flight over and under low stratus. In K. Schmidt-Koenig and W.T. Keeton (eds.), *Animal Migration, Navigation, and Homing* (pp. 162–170). Springer, Berlin.

214 Hagstrum, J.T. (2019) A reinterpretation of "Homing pigeons' flight over and under low stratus" based on atmospheric propagation modeling of infrasonic navigational cues. *Journal of Comparative Physiology A*, 205, 67–78.

215 Normally as one ascends through the atmosphere the temperature drops. If there is an inversion the temperature rises with height. This is particularly dangerous for hot air balloons which rely on a temperature differential for flight. If they fly into hot rising air, they can dramatically fall. The same is true if one flies into a thermal.

216 Freeman, A.R. and Hare, J.F. (2015) Infrasound in mating displays: A peacock's tale. *Animal Behaviour*, 102, 241–250.

217 McComb, K., Reby, D., Baker, L., Moss, C. and Sayialel, S. (2003) Long-distance communication of acoustic cues to social identity in African elephants. *Animal Behaviour*, 65, 317–329.

218 Georgia, Department of Natural Resources: https://georgiawildlife.com/out-my-backdoor-can-birds-predict-severe-weather? (Accessed 08/08/22)

219 "Uncle Offa" (1991) *Natural Weather Wisdom*. The Self Publishing Association Ltd., Worcs, UK.

220 The Times: www.thetimes.co.uk/article/climbers-killed-as-italian-glacier-crumbles-in-heatwave-bfzgk60gd?shareToken=d0f83e2b2049c0532de3b7fb9ea769c2 (Accessed 05/07/22)

221 The Guardian: www.theguardian.com/world/2022/jul/26/melting-alps-theodul-glacier-switzerland-italy-border-shifts (Accessed 03/08/22)

Predicting earthquakes, volcanic eruptions and tsunami

4.1 Introductory story

The mountain was looming over her. It was an ever-present threat, despite not having erupted for years. Although not designated as extinct, she couldn't remember the last time this dormant volcano had caused any trouble.

The young woman looked around for her herd of goats, but they were gone. The ground was mainly spree, comprising small rocks through which the odd daisy was trying to grab enough light to survive. But of the goats there was no sign.

The woman climbed higher, hoping to find them over the next ridge. It was a hot and sunny day, and as she struggled to get enough grip on the loose ground, she started to sweat. This wasn't fair. All she wanted was to go home. Why did the pesky goats have to misbehave on this day, of all days? She had a date that evening and needed to get back for a quick bath and to get ready for the evening, not be rushing around like a mad woman. At the ridge there were no goats, or any sign of them. At a loss, she headed back down the mountain to Marco's farm to find out if he had seen them. Marco was an old man and had been farming on the side of the volcano for over 40 years. There was little that escaped his notice. If the goats were near his farm, he would know.

Ten minutes later she was striding across the farmyard, but Marco was already there. "We need to go," he greeted her.

"But I've just got here!" the woman exclaimed. "You seen my goats?"

"Yes, and that's the point. They ran by ten minutes ago. They were heading down the valley."

"Oh, thanks." She sighed. "I'd better chase after them."

"No, leave them. There's an eruption coming!"

The woman looked back at the volcano. The rim of the cone was clearly visible from Marco's farm, crisp and clear against the bright blue sky. There was no sign of activity.

"Looks fine to me," she shrugged.

"Trust me. If not me, trust the goats. Something's going to happen," Marco said.

As if on cue, the woman watched as a small plume of ash erupted from the left side of the cone. "Is that-?"

"Don't just stand there. Jump in the truck. We're out of here. If she blows, we could be covered in seconds."

"And you knew this was going to happen?" The woman asked the old man.

"You need to watch and learn. The animals do. Why do you think they left in such a hurry? Start praying that we have farms to return to."

The woman took one last look at the volcano. The ash plumb was getting larger. She jumped in Maroc's truck and they both chased down the valley after the goats.

4.2 Earthquakes and other rumblings – an introduction

The prediction of the weather, whether being carried out using modern instrumentation or by observing animals, is relatively easy to comprehend. It is part of everyday life for millions of people, whether it is on the news or signalled by a flock of birds. However, more extreme environmental events are harder to understand. These include earthquakes, tsunami and volcanic eruptions. Although there is modern equipment to aid in forecasting their occurrence, such events are still very hard, perhaps impossible, to predict. Can animals come to our assistance?

Earthquakes are amazingly frequent. The National Earthquake Information Center estimates that there are 12,000–14,000 earthquakes each year,[1] but of course, most of these would be of little or no consequence and may not even be felt. The magnitude of earthquakes is measured on the Richter scale, which is relatively commonly known. This is not a linear scale, but each point of increase is a factor of 10 (i.e., a logarithmic scale),[2] so an earthquake of magnitude 5 is 10 times stronger than that of magnitude 4, and so on. Therefore, at the top of the scale (beyond 9) the earthquake is extremely violent and may lead to total destruction of the environment. The Richter scale was superseded by the moment magnitude scale (MMS), defined by Thomas C. Hanks and Hiroo Kanamori in 1979.[3] This is also a logarithmic scale. The most violent earthquake recorded was the Valdivia Earthquake, in Chile in 1960. This measured 9.5 on the scale, with the next most violent being the Great Alaska Earthquake, in 1964, registering in at 9.2.[4]

The Richter scale was named after Charles Francis Richter (1900–1985), who published his ideas in an extremely long paper in the *Bulletin of the Seismological Society of America*.[5] Although published in 1935, the endnote of the paper says that it was from the Carnegie Institute of Washington Seismological Research Pasadena, California, 8th June 1934. Richter was born in Ohio, went on to study at Stanford University and in 1952 became professor of seismology at California Institute of Technology.[6] His magnitude scale and hence his name have almost become synonymous with earthquakes and are still referred to today.

Hanks and Kanamori[7] appear to be worried about the upper limits of the measurement scales being used, and tabulated the measurement of earthquakes known to have been recorded in California, such as the 1906 quake in San Francisco. However, even relatively recently, the accuracy and usefulness of different scales is still being discussed.[8]

4.3 Instrumentation which has been developed to detect movements of the ground

The shake felt during an earthquake is measured with a seismometer (seismograph if it records).[9] An instrument which could measure such a phenomenon was probably first created by Zhang Heng (78–139).[10] He was a Chinese mathematician and astronomer. His instrument was called the *houfeng didongy yi*, which in English translates to "the instrument for inquiry into the fluid and earth movement". It is not certain what this instrument looked like. There was probably only one example ever made, and shortly after his death it was moved to Zhang Heng's tomb.[11] Unfortunately, during a raid of the Mongolian army the seismometer was destroyed. Remnants of it were excavated by archaeologists, but it is still not clear what it looked like or how it worked. Some images show a pendulum, whilst it may have used a flow of water. It was probably cylindrical with eight dragons' heads around the top. Each dragon contained a ball in its mouth and directly below were eight frogs with their mouths open. The shaking of the instrument would cause the balls to fall from the dragons into the frogs, creating a noise and thus sounding an alert. It was not the most accurate of machines, and you had to be near enough to it to hear it. If you returned a few hours later and saw that the balls had dropped, it would be too late. However, it was certainly revolutionary for its time. It was said to have detected an earthquake on 1st March 138 AD which was 500 km away and destroyed the city of Longxi (Western Gansu Province). However, it was also purported to be a very difficult instrument to maintain and on Zhang

A stamp printed in China shows Zhang Chang Heng (78–139), astronomer, Portraits of Scientists, 1955.
(Image from Shutterstock, but made black and white.)

Heng's death it fell into disrepair and was never used again. Although novel for its period, and certainly showing that there was a requirement for such an instrument, as well as the forward thinking of Zhang Heng, it was not very good as a predictive instrument as it relied on the earthquake actually happening. On the other hand, if it could sense tremors hundreds of kilometres away, it would have given some early warning of major tremors, if only on a short timescale. Perhaps enough warning to leave a building which otherwise could crush you to death.

The next development in the seismograph was probably created by Luigi Palmieri (1807–1896) in 1855. He was a scientist from Benevento, Italy. His instrument contained mercury in U-shaped tubes, each one pointing in a different direction. When the instrument shook, the mercury caused an electrical circuit which stopped a clock and started a recording drum.

Therefore, it could record both exactly when the earthquake happened but also the strength of the shaking. He was prompted to invent this instrument when he was studying volcanoes and realised that the eruptions and earthquakes (or at least the ground shaking) were connected. This was particularly focused on Mount Vesuvius, Italy, and was reported in the *Aunali dell' osservatorio Vesuviano* (1869–1873).[12]

An instrument to measure quakes which had an inverted pendulum was used in Perthshire, Scotland in 1840, even though earthquakes in Scotland are relatively rare, and destructive ones hardly ever happen.[13] The first modern seismometer was invented by another Italian scientist, Filippo Cecchi (1822–1887). His instrument also was based on a pendulum, and could record the strength and duration of a quake. Cecchi was born in Tuscany and from 1872 to 1887 was the director of the *Osservatorio Ximeniano* in Florence, Italy, and he became the vice president of the Italian Meteorological Society.[14]

More modern instruments are based on electromagnetic principles, involving pendulums. It is important to measure the movement of the ground in three dimensions (two dimensions horizontally and also up and down[15]). Modern instruments can also amplify the signal so are extremely

Eruption of Vesuvius 1760–1761: Campi Phlegraei – Sir William Hamilton.
(Available at the Wellcome Collection with a Public Domain Mark. This image has been made black and white.)

sensitive. Other variations include the strain seismometer, suggested by a seismologist, Hugo Benioff (1899–1968),[16] in 1935. He was an American, born in Los Angeles, and worked at the California Institute of Technology. His father was from Russia whilst his mother was from Sweden. He graduated in 1920 from Pomona College. Interestingly, he took internships at Mount Wilson during his undergraduate studies and therefore it would appear he would have been there at the same time that Harlow Shapley was pondering his ants (Chapter 2). Perhaps they even met, but I have seen no evidence of this.

Seismographs, and related instrumentation, will very accurately measure the intensity of an earthquake and its duration. They are not only limited to this but will measure any shaking of the ground. This might be natural, such as caused by an earthquake or volcano, but could be from anthropogenic activity too, such as explosions (nuclear and non-nuclear), and fracking.[17] Seismometers have even been used on the moon. The first instrument used there was solar powered and contained four seismometers. It was placed on the moon by the Apollo 11 mission. Further instruments were left up there by Apollo 12, 14, 15 and 16, and they returned data back to Earth until September 1977.[18] One assumes that they are still there, just no longer working. All sorts of things have been abandoned there, including personal items such as photographs, but also equipment which is too expensive to retrieve, such as the moon buggy.[19,20] It has been estimated that there might be 400,000 pounds of items left up there. There are even golf balls on the lunar surface.

Although today's measurement of seismic activity is extremely accurate, it will not predict that a quake will happen. It might predict a second, perhaps assumed to be a weaker quake, but it is not likely to predict the initial occurrence. So, can animals do any better?

4.4   Why predicting geophysical events is important

Predicting earthquakes, volcanic eruptions and tsunami is vitally important. There are several regions of the world where imminent disaster is predicted, but when will it happen? Will we get enough warning? If not, the consequences are going to be truly terrible.

There are some startling examples of possible impending disasters around the world. One of the most famous is the San Andreas Fault.[21] The surface of the world is composed of tectonic plates, and the places where these meet are often referred to as fault lines. The plates move, and when they do, they can create earthquakes which are felt on the ground. The San Andreas Fault is the border between the Pacific plate and the North

American plate, and runs down the west coast of North America, from Cape Mendocino to the border with Mexico. Along that line are numerous major cities and towns. Two major earthquakes on the fault were the 1906 event in San Francisco which was 7.8 in magnitude, and one in Loma Prieta in 1989, which was 6.9 in magnitude.[22] In the 1906 earthquake much of the devastation was caused by fire in the aftermath of the shake itself. It was estimated that 28,000 buildings were destroyed with a value in the region of $350 million.[23] An estimated 3,000 people lost their lives, with 250,000 left homeless as swathes of the city were flattened. If the San Andreas Fault were to cause another major earthquake, the modern-day destruction could be far worse. Despite the 1906 quake, it is said that there has not been a major event in southern California since 1857 and, with the fault having a significant shift every 150 years (approximately), people talk about it being "overdue". Some commentators[24] think that the next "big one" will be at the southern end (estimated to be 7.8 magnitude). If so, it could cause the death of approximately 1,800 people with more than 50,000 injured. With the area having a major presence of oil and gas industries, the potential for fires is huge. The problem is that it is not "if" this event will happen, rather "when", and this is not easy to predict. Early warning systems rely on picking up the shaking and therefore at the epicentre there is virtually no warning. The farther away you are, the more warning you may get, but that may still be a matter of seconds. Perhaps this amount of time may be long enough to shut off a gas supply or stop a train (which may be approaching a bridge for example), but not long enough to pack the car and leave.[25]

To emphasise the destruction an earthquake can cause, it is worth pausing over the 1755 quake which hit Lisbon, Portugal. The city was struck on 1st November that year, and it has been estimated that the quake had a magnitude of 8.5–9.0 on the Richter scale. The earthquake resulted in an estimated 60,000–75,000 deaths in the city with approximately 12,000 buildings being destroyed.[26] Many of these were churches and, as it was All Saints' Day, many people were in those churches praying as they fell. Fire was also a problem, partly because of the use of candles and because of the construction of the buildings, as well as the decorations placed on them for the celebrations. The earthquake also caused a tsunami which was estimated to be 20 metres high when it crashed into Cadiz, Spain. This quake was propagated to the west across the Atlantic too. Interestingly, in the National Tile Museum (Museu Nacional do Azulejo), Lisbon, is a frieze constructed in tiles which depicts what Lisbon was like before the "Great Earthquake". It is amazingly complete, at 23 metres long, showing what 14 km of coastline looked like. This frieze was thought to have been created by a Spanish craftsman, Gabriel del Barco (b. 1648).[27] Pictures of it can be found on the internet, for example at the Portugal Tourism Guide.[28] Sadly, much of what can be seen in this frieze was destroyed by the earthquake in 1755.

Another "overdue" world event is the eruption of Mount Vesuvius in Italy. This was the focus of the concerns of Luigi Palmieri in the 19th Century, as discussed earlier. The major eruption of 79 AD hit Pompeii[29] and wiped the civilisation from the face of the Earth. Approximately 2,000 people died, but it has been estimated that the total death toll of the eruption might have been as high as 16,000 people, and the settlement was abandoned. As the eruption continued the city was hit by a pyroclastic surge, a fast-moving wave of gas and ash, impossible to escape from. Volcanoes release many toxic gases too, some sulfur[30]-based, such as hydrogen sulfide (H_2S). This is very poisonous[31] and will contribute to the deaths seen. A major earthquake had hit the same area in 63 AD, which should have been a warning, but it was not understood in those times what the significance of this was. Pompeii was not the only community hit either, with major losses at Herculaneum, Torre Annunziata and Stabiae. The area has now been extensively excavated and is a significant archaeological site, having UNESCO World Heritage status since 1997.[32] Will it happen again? Yes, we can assume so, since it is still an active volcano. Even though the rim of the crater can be visited,[33] it will one day erupt. In his paper to the Royal Society, *X. On the phænomena of volcanoes*, read on 20th March 1828, Sir Humphry Davy states:[34]

A depiction of the Lisbon earthquake of 1755. Note the disruption of the sea and the damage that is shown.
(Sourced from Wikimedia Commons. This image has been made black and white.)

Tile frieze at the National Tile Museum (Museu Nacional do Azulejo) in Lisbon, showing what the city looked like before 1755, that is, before the earthquake struck.
(Photo by Dr Sally-Ann Kitts. This image has been made black and white.)

Roman road in the ruins of Pompeii.
(Photo by Matt Jones on Unsplash. This image has been made black and white.)

The active volcano on which I have made my observations is Vesuvius; and there probably does not exist another so admirably fitted for the purpose;

And he goes on to say:

I had made several observations on Vesuvius in the springs of 1814 and 1815, which I shall refer to on a future occasion in these pages; but it was in December 1819, and January and February 1820, that the volcano offered the most favourable opportunity for investigation. On my arrival at Naples, Dec. 4, I found that there had been a small eruption a few days before, and that a stream of lava was flowing with considerable activity from an aperture in the mountain a little below the crater.

In the 19th Century the volcano was clearly active, as it is today (although it is officially designated by some as dormant[35]). The most recent significant eruptions were from 1913 to 1944,[36] but it will erupt again. In its shadow is the city of Naples, which in 2021 had a population of 2,182,885.[37] Vesuvius is thought to erupt on average every 46 years, so there is an eruption "due".[38] Again, such phrasing highlights that no one knows when this will happen. Early warning systems are in place,[39] but this article by Gasparini et al. states: "this area has the highest volcanic risk in Europe and one of the highest in the world." Clearly this will happen, but when and how much warning can there be before it does so? Naples has an evacuation plan for 700,000 residents who may be affected[40] but this will take a long time to implement. The plans use a 72-hour period, with a 12-hour safety margin. However, it seems unlikely that the mountain will give the authorities that much warning.

As I write this, there is a significant volcano eruption causing havoc in Tenerife.[41] A nuisance to anyone wishing to go there on holiday, it is a major issue for those who live there. Like many eruptions, the spewing of lava is accompanied by earthquakes and there is a risk of a tsunami[42] as well as pyroclastic surges. It is not good news for the islanders.

Of course, not all volcanoes erupt with devastating consequences. Volcanoes such as Mount Etna, on Sicily, seem to rumble all the time and have become a tourist attraction in their own right. Etna even has its own cable car.[43] This is fine until the volcano gets more violent and does significant damage, which is always possible at any time, and not easy to predict. In 1669 Etna erupted, destroying 15 villages and part of Catania,[44] a city port on the east coast of the island. Today this city has a population of over 300,000 people. Even rumbling volcanoes can't be ignored.

In 2004 the Indian coast was hit by a tsunami.[45] A tsunami is defined as "a series of waves caused by earthquakes or undersea volcanic eruptions".[46] On 26th December 2004, a 9.1 magnitude earthquake hit off the coast of Sumatra Island, Indonesia. The fault line between the Indian and Australian tectonic plates sifted across a 900-mile stretch. This quake caused the ocean floor to be thrust upwards by approximately 40 metres, creating a massive tsunami. Then, 20 minutes later the 100 feet high waves hit Banda Aceh, in Indonesia, and it was estimated that more than 100,000 people were killed.[47] The devastation didn't stop there either. The waves rolled on and hit the coasts of Thailand, India and Sri Lanka. It was even reported that the waves reached, and caused death and damage in, South Africa. That is 5,000 miles away, and it took the waves eight hours to get there. In total, the death toll was estimated to be in the region of 230,000 people. This is both incredible and devastating. It highlights the power of rumblings of the Earth's crust. More recently, there are warnings that areas of the Mediterranean may be in danger of being hit by a tsunami. This in particular may involve Marseille, Alexandria and Istanbul,[48] so a wide area across the Mediterranean. The threat of such natural disasters never goes away.

The manner in which the waves of a tsunami can be propagated is depicted in a representation of a tsunami by Von Hochstetter which was drawn after the 1868 Arica earthquake. It suggests that the waves can indeed cover vast distances, as in this drawing the waves are shown to be able to travel from South America to New Zealand, across the whole Pacific Ocean. Christian Gottlieb Ferdinand Ritter von Hochstetter (1829–1884) was a German-Austrian geologist.

I have a friend from *Chennai*,[49] a major city on the south-eastern coast of India, and at the time of the December 2004 tsunami, she was back home, visiting her family for the Christmas break, as a lot of people were doing. She told me that she had been walking on the beach just before the wave hit.[50] She gave the distinct impression that she was lucky to have survived. I am sure others had friends and colleagues who were not so lucky. Such disasters hit people around the globe, not only those in the immediate vicinity.

There is considerable effort in trying to forewarn against events such as tsunami. Tsunami warnings can be gleaned from several sources, including the International Tsunami Information Center,[51] which lists several warning stations around the world including the US Tsunami Warning System,[52] Northwest Pacific Tsunami Advisory Center (NWPTAC[53]) and the South China Sea Tsunami Advisory Center[54] (SCATAC). Levels of threat are "warning", "advisory" or "watch". A warning is given when widespread inundation is imminent, expected or occurring. A "watch" is based on known seismic activity but does not mean that a tsunami is known to occur.

In a similar manner there are several earthquake and volcanic eruption warning sites. Earthquake warnings can be found at sources such as Shake Alert for the west coast of USA. It does not predict an earthquake. According to their own information: "ShakeAlert is not earthquake prediction, rather a ShakeAlert Message indicates that an earthquake has begun and shaking is imminent."[55] Other systems include Earthquake Warning California.[56] Again, this relies on sensors picking up earth movements so the earthquake is already happening. Elsewhere in the world, Japan has an earthquake early warning (EEW) system,[57] also relying on the earthquake happening. In China there is the development of early warning network,[58] which is projected to be completed by June 2023, but has already been evaluated in the academic literature,[59] which concludes that the system is ready and that they can "launch the trial operation in the pilot areas for public early warning services". An informative website about the EEW in Italy can also be accessed.[60] All the information on these systems stress that it is not prediction but warning of imminent danger as the shakes progress.

For volcanic eruptions, warning systems are also in place. In the USA, the USGS says, "Roughly half of the Nation's 169 young volcanoes are dangerous because of the manner in which they erupt" and the institute is involved in volcano monitoring.[61] In Japan, the Japan Meteorological Agency issues alerts on a five-point scale.[62] In Italy, monitoring of infrasound was found to be useful for eruption prediction,[63] and this could be used for an automated system. Using Mount Etna as a model system the volcano could be monitored from hundreds of kilometres away, suggesting that a global array of sensors could be used to monitor a range of volcanoes simultaneously. For the Pacific region, Meteologix[64] uses false colour imaging from satellites to monitor volcanic activity from ash clouds.

4.5 Animals and how they can be used to predict the geophysical events

With so many systems and the amount of effort being focused into the monitoring of earthquakes, volcanoes and tsunami, as discussed earlier, can animals help to prevent such disasters?

This is certainly not a new idea. Rats, weasels, centipedes and snakes were reported to have moved to safety before an earthquake in Greece in 373 BC.[65] This has subsequently been used by others to indicate the start of this type of observation.[66]

In 1982, a physical chemist called Helmut Tributsch looked at the evidence and wrote a book entitled *When the Snakes Awake: Animals and*

Earthquake Prediction.[67] This was partly inspired by the destruction of his own village in northern Italy by the Friuli earthquake in 1976. He went on to look at 77 other earthquakes and collated anecdotal evidence of whether animals were able to predict what was about to happen. He suggested that the atmosphere had an increase in electrostatically charged particles which may have accounted for what was observed.

"Uncle Offa" also lists, in a brief section, how animals can be used to predict earthquakes.[68] He says that his information comes from Chinese folklore, and lists such things as "Horses run wild . . .", "Cats carry their kittens out of doors," and "Snakes leave their nests." Other examples from here are listed in Table 1. Offa goes on to say that the animals are sensitive to an increase in ions on the atmosphere, but does not say how the animals perceive these ions.

So, what is the evidence and the scientific underpinning of such animal observations prior to earthquakes and other dramatic events?

In a list of animals which can be useful for predicting changes in weather, elephants are listed as being able to forecast earthquakes.[69] It was suggested that this is either because they have good hearing or because they can feel vibrations. Either way, this is hardly a long-range warning, but rather another indication of imminent shaking. It seems as good as those instruments that warn of an earthquake based on the event having already started. If elephants can be observed for earthquake warning, what about other animals?

In a study of farm animals, i.e., cows, dogs and sheep, which were close to the epicentre of the 2016 Norcia, Italy, earthquake (magnitude 6.6), a bio-logging survey using motion sensors on nylon harnesses suggested that animals kept indoors, such as in a stable, *could* predict the earthquakes but those kept outside on pastures could not. Prediction times were up to 20 hours before the tremors. The researchers looked at a range of seismic activity, from 0.4 to 6.6 magnitude, encompassing over 18,000 earthquakes during 2016 and 2017 and found that animals often had anticipatory behaviour. They suggested that using such bio-logging technology on animals would be a useful predictor of seismic activity.[70] If prediction times were indeed a matter of several hours, this would be a significant warning with a useful timeframe, enabling evacuation or the shutting down of services which could cause further devastation, such as fuel supplies. It still does not come close to the 72 hours planned for in Naples.

We met the observations of seagulls in Chapter 3 when considering the weather, and here they are back again as indicators of earthquakes:

> Picking up on infrasonic pulses, seagulls around the world have flown inland a day or two before major earthquakes, sometimes as much as five kilometers, or several miles.[71]

If this is true, it is really remarkable, as we are told gulls can give a day or two's warning. This implies that these birds are more sensitive to some environmental changes (sonic waves?) long before seismometers are picking up any rumbling on the surface. Two days would mean twice as long as the warning from the farm animals in Italy, and would give a much better evacuation window.

As one might suspect, there is a lot of scepticism about the ability of animals to be able to predict earthquakes. A report by the U.S. Geological Survey stated: "Changes in animal behavior cannot be used to predict earthquakes."[72] On the other hand there are others who think that animals can be a good indicator, so what is the evidence?

There are reports of pets disappearing just before an earthquake and even this being used to predict an earthquake in San Francisco in 1989.[73] However, others say that there is no statistical evidence for this. Using qualitative observations and quantitative evidence, a quite extensive study was published.[74] The study methods were to:

> compare the daily quantities of lost and found pets reported in the San Jose Mercury News with dates of earthquake events in the San Francisco Bay area over a statistically significant period of time – a three-year period from January 1, 1983 to December 31, 1985.

The author, R.B. Schaal, who was a geologist at University of California, Davis, looked at 41,717 daily reports of missing pets and compared this against the records of 224 earthquakes (magnitude 2.5 or greater). The conclusion was rather definitive:

> This study shows that a significant positive correlation does not exist between the behavior of pets . . . and the occurrence of earthquakes . . . no scheme seems possible to predict earthquakes using newspaper reports of missing pets.

[I added the gaps for clarity]

However, there might be other evidence that does show that animals can be used as earthquake predictors. In 2010, R.A. Grant and T. Halliday suggested that toads were a good predictor, in particular *Bufo bufo*.[75] They looked at the behaviour of these animals for 29 days, around the earthquake at L'Aquila, Italy, in April of the previous year. They monitored the toads for nine days before and 20 days after. Even though they were 74 km from the epicentre, what they noticed was that there was a "dramatic change in behaviour 5 days before" the quake. The toads stopped their spawning and only resumed normal behaviour several days after the earthquake. Of particular pertinence here was the comment:

> It is unclear what environmental stimuli the toads were responding to so far in advance of the EQ, but reduced toad activity coincides with pre-seismic perturbations in the ionosphere, detected by very low frequency (VLF) radio sounding.

Here, not only were the earthquakes being predicted, but the authors postulated a reason. There was something tangible which might be sensed by the animals, and this was not relying on the earth shaking at the time.

During an earthquake in Eilat, Israel (magnitude 7.2) birds were observed taking to the air prior to the tremors. In a paper from 1997,[76] the author notes that although the earthquake was reported at 6.16 am:

> I first observed unusual behaviour at about 06:00 hrs when flocks of several avian species flew north. Also, a flock of about 50 Grey Herons (*Ardea cinerea*), that roosted on the salt pans, became increasingly restless and eventually took to the air at 06:08 hrs.

Many other birds reacted, including gulls (*Larus* spp.), kingfisher (*Ceryle rudis* and *Alcedo atthis*) and cormorant (*Phalacrocoras carbo*), but smaller birds appeared not to be affected. The observations continued, to see how the birds would react as the tremors continued. It may just be a sensitivity issue, but the abstract of the paper concludes as such:

> One of the interesting and consistent observations was that the birds showed the first signs of restlessnes about 30–60 seconds prior to the human observers ability to feel the tremors.

[spelling mistakes here were in the original]

It is interesting that the larger birds here were the ones that responded, and this may be because they are more sensitive. After all we saw elephants being particularly sensitive, although us humans, being even larger than the birds, did not seem to be able to sense anything anticipatory (perhaps the people were wearing thick-soled shoes or boots?). However, in discussing this paper, Grant and Halliday say that they think this may be a genuine anticipatory reaction by the birds. Interestingly, it was also noted that the authors could only find four previous papers on the reaction of birds to earthquakes. These were written in Russian and Chinese, and all published in 1993 or 1994 (this one was 1997).

In Japan, before the Kobe earthquake of 1995, it was reported that the circadian rhythms[77] of mice was disrupted during the day before the tremors happened.[78] The mice activities were measured using an infra-red sensor and were found to be increased during the sleep and active times. Four cages were used for the mice and the authors noted that one was broken by the earthquake, so it was clearly a significant event (magnitude 7.3). Similar data were reported during the earthquake in China in 2008 (in Wenchuan county, magnitude 8.0). Eight mice were monitored and their activity overall decreased and they lost their circadian rhythm-induced activities.[79] This was observed on day 3 before the earthquake and the effects lasted for six days. The authors suggested that the animals were sensing changes in geomagnetic intensity. In rats, during the Wenchuan earthquake, just before it happened, researchers noted that the animals had effects on insulin-induced signalling, which went along with a decrease in glucose uptake in skeletal muscle (SkM) and adipose tissue (AT: that is white fat tissue).[80] It was concluded this was a stress response, manifesting in increased food intake and increased insulin resistance, but this was before the earthquake. This study, it was pointed out, was undertaken 1000 km from the epicentre, but it was also noted that the earthquake could still be felt. It was also highlighted that the timing of this study was a coincidence, as they were intending to measure these physiological parameters in these animals anyway – serendipity is a wonderful thing. However, such studies do suggest that animals, and in this case mammals, seem to be sensitive to something changing before an earthquake happens.

However, in a study of ants (*Messor pergandei*) there were no evidence of altered behaviour around the time of the Landers earthquake in the Mojave Desert, USA.[81] This was rated with a magnitude of 7.4 on the Richter scale, so was a significant quake. The authors measured a range of parameters before, during and after the earthquake. These included trail speed of the ants, worker mass distributions, rates of aerobic catabolism and temperature at the ants. Their conclusion was: "anecdotal accounts of the effects of earthquakes or their precursors on insect behavior should be interpreted with caution." I can't help thinking that Shapley would have been disappointed by this.

A study was published by Bhargava *et al.* in 2009, but with a focus on how earthquake detection might be of pragmatic use in India.[82] Here they report that the Haicheng earthquake (central Liaoning province, China) (magnitude 7.3), which happened on 4th February 1975, was predicted in mid-December 1974 by observing animal behaviour. Indicators included the appearance of rats, and snakes which came out of hibernation. There were earthquakes in that December. And then, during the first days of February, unusual behaviour of a range of larger animals was noted, including cows, horses, dogs and pigs. In his blog,[83] Steve Smith, a professor of history at the University of Essex, UK, says that:

> From December onwards, people began to report dazed rats and snakes that appeared 'frozen' to the roads. From February there were numerous reports of cows and horses appearing restless, of chickens refusing to enter their coops, and of domestic geese taking flight.

The earthquake struck on the 4th February.

4.6 Biochemical discussion – what animals may be sensing to predict movements of the earth

If animals are truly predicting an earthquake before it happens, they must be sensing something. Some of the preceding discussions mention several explanations. These include simply that animals are better at feeling vibrations than humans and that many species have better hearing. This may include sensing very low frequency (VLF) radio sound, or infrasonic pulses.

The shock waves emanating from an earthquake are not simple, but the tremors travel as compressional (P) waves and shear (S) waves.[84] The P waves travel much faster but are of less significance when it comes to structural damage. The slower S waves, which cannot pass through the Earth's core,[85] are what cause the destruction. Animals may be able to sense the P waves before the S waves arrive. Obviously the farther this would be from the epicentre the less likely that they will be felt and also the longer the prediction time. This may account for why animals such as elephants can be used as a predictor: they can sense the P wave before the S arrives, so giving a short-term warning. It may be what other animals, such as birds on the ground, are also sensing.

Other explanations include an increase in electrostatically charged particles and geomagnetic intensity in the atmosphere. All these phenomena need to be propagated before the earthquake can be felt and then sensed by the animal.

But why should animals have become sensitive to such phenomena?

A comprehensive treatise on the possibility of animals in predicting earthquakes was published by Joseph L. Kirschvink in 2000.[86] He was not only looking for examples, but discussed whether it is possible from an evolutionary point of view, and what the animals may be sensing. He concludes that it is possible that such sensing systems have evolved. Evolution was possible, partly because there is lots of seismic activity, not just on the edge of tectonic plates, and also because animals for which earthquakes are likely to increase death rates, such as burrowing animals, have been around a very long time.

> Thus, mammals in particular have had over 250 million years in which to refine their seismic escape response and link it via exaptation to additional sensory signals.

As this paper points out, many observations of animal behaviour and seismic activity are serendipitous, as significant earthquakes are so hard to predict it is extremely difficult to set up an experiment and sit around waiting. On the other hand, there are a selection of things which could be being sensed and therefore tested for. These include ground tilting and humidity changes. The former is theoretically possible but there seems to be a dearth of data in animals, even if some are thought to be much more sensitive to this than humans. The detection of humidity before seismic activity is also possible, especially if ground water rises, and some animals such as insects and spiders may be particularly sensitive to humidity changes. So, this may be a potential way animals sense earthquakes before they happen, but as pointed out, it is unlikely to work well in regions of the world where humidity is already, and rather constantly, high such as in Japan. Kirschvink goes on to discuss if electrical currents may be important, but he points out that although aquatic animals are often sensitive to electrical changes, this is less likely in land animals. To emphasise how aquatic animals may use such systems he says:

> In the elasmobranch fish (sharks and rays), a specialized receptor system in the ampullae of Lorenzini has, in fact, reached the thermal noise limit with the ability to perceive nanovolt changes in electrical fields (Kalmijn, 1974[87]); these are comparable to the voltage of a flashlight battery applied across the Atlantic Ocean.

That seems truly incredible.

The last factor which Kirschvink discusses as being responsible for the behaviour changes seen in some animals is the alteration of the magnetic field. Like all environmental factors which can be used to explain what is going on there are two underlying questions. Firstly, do these characteristics alter before the earthquake and so can be used as predictor? Secondly, can animals actually sense them? Certainly, for magnetic fields the first is possible:

> a significant elevation in magnetic activity in the .01–5 Hz frequency range starting about two weeks prior to the earthquake, with peak amplitudes in the 1–3 nT range. About three hours before the event, however, the largest signals exceeded the dynamic range of the instrument.

To answer the second question Kirschvink quotes several papers which say that many organisms can detect magnetic fields using a ferrimagnetic mineral (magnetite: Fe_3O_4). This was apparently first found in bacteria (so called magnetotactic bacteria),[88] but similar systems have been found in fish such as tuna,[89] birds such as pigeons[90] and insects such as honeybees.[91] Alternatively, animals may contain proteins termed cryptochromes. These act as light-dependent magnetoreceptors.[92] It is argued that some animals, particularly if they primarily live in the dark and cannot therefore use sunlight as a cue, will use fluctuations in the magnetic field to time their circadian rhythms (a phenomena altered in rodents, see earlier), whilst some animals may have evolved such sensing to aid in navigation.[93,94]

Howard T. Odum, in 1949, posited:

> Is the superior navigation of birds possible because of their possible ability to orient to a Coriolis, magnetic, or other geophysical field of force in addition to keen powers of visual reference.[95]

He goes on to rule out the effect of the Coriolis Force[96] in bird magnetism but suggests that there seemed to be positive effects of magnetism, although that the work needs repeating. To investigate this further, researchers have attached magnets to birds and then watched what happened. William Keeton, at Cornell University did just this with pigeons.[97] He glued magnets (255 gauss) to the backs of the birds, just at the base of their necks. The field strength of the magnets at each bird's head was estimated to be about 0.45 gauss, while the control group of birds carried a brass bar, so that the influence of gluing and weight could be ruled out. Birds were released 27–50 km from their lofts and then he observed. The birds carrying the magnets were often disorientated. Interestingly, this seemed to happen only when the weather was overcast, and not during a sunny day, suggesting that the birds were using multiple cues for navigation. Yasuo Harada, at Hiroshima University, sectioned the lagenal nerves of pigeons, and used magnets. Treated birds became lost or disorientated, unlike the control group, and Harada concluded the lagena is the organ responsible for sensing that magnetic fields in birds.[98] Using synchrotron X-ray fluorescence analysis he went on to show that iron is contained in the lagenal otoliths of the birds, and also in sea fish.[99]

A study to investigate magnetic effects on birds was also carried out by Massa *et al.*[100] They chose Cory's shearwaters (*Calonectris diomedea*) as their experimental animals, but again used magnets and brass bars as a control. The magnets were attached to the back of the birds, as well as the base of their necks and upper part of each wing. There was some magnet loss during flight noted. However, the results showed that the magnets had no effect, and the homing ability of all the birds appeared to be the same. The authors concluded that their data did not support any effects of magnetic forces on the homing of Cory's shearwaters. Clearly, if there are effects in pigeons, this cannot be automatically extrapolated to other species. This conclusion was supported by work with the Black-browed Albatross (*Diomedea melanophris*).[101] Their foraging activity was determined in birds which had magnets, or dummies, attached and no difference was found. The authors say that they could not rule out that the birds were using magnetic cues for navigation back to their nests but that it was unlikely to be a major factor. Such data were confirmed by others. The foraging routes of waved albatrosses (*Phoebastria irrorate*: sometimes referred to as the Galapagos albatross) between the Galapagos Islands and feeding grounds off Peru were reported.[102] As with other studies, birds had magnets attached, or had dummy metal bars attached. The routes taken by the birds were tracked using satellite telemetry. The magnet had no effect on these free-flying birds, but the authors said that they could not rule out that the birds were using some magnetic sensing in their head that was not known about. Similar results were found for Green turtles (*Chelonia mydas*) when they were monitored, with and without magnets attached, as they moved between their nesting beaches on Ascension Island and their feeding grounds in Brazil.[103]

Very recently it has been reported that magnetic fields alter the levels of reactive oxygen species (ROS) in the hippocampus of mice.[104] ROS were first brought to the attention of biologists as they are involved in warding off pathogens, in what was dubbed the respiratory burst.[105] ROS include molecules such as hydrogen peroxide (H_2O_2: often used as bleach) so this was not controversial. What became more surprising was later discoveries that these relatively reactive molecules (it is in the name after all) are used as instrumental control substances in cells regulating, for example, gene expression. If too many ROS accumulate cells are said to be under oxidative stress, this can lead to a range of diseases, including neurodegenerative and cancer. Therefore, the levels of ROS in biological systems are a balance. Low levels are useful; high levels are harmful. It appears that magnetic forces may have an influence on this. In a similar manner, others have shown that nitric oxide (NO) metabolism can be affected by magnetic fields.[106] Like the ROS, superoxide (O_2^-), NO is also a radical[107] (and a gas at atmospheric pressure), which is instrumental in the control of numerous cellular functions and is implicated in a range of disease states. It seems as though the topics of magnetic effects in biological systems and 'magneto-medicine' are gaining traction in the scientific literature.[108]

Therefore, there is evidence that some animals can be affected by magnetic fields, but several experiments posited to show that magnets can affect animal behaviour failed. Perhaps in some animals, but not all, magnetic sensors in cells may be triggered before an earthquake and alert some animals to possible impending doom. It does seem as though changes in the geomagnetic field can precede the quake and be sensed by some animals.

In a recent report by Michael Garstang and Michael C. Kelley[109] they state: "No predictions of earthquakes have withstood careful scrutiny." However, that is not the end of their paper, and they examine the sounds which may be heard by animals before an earthquake, to try to predict what the animals may be able to sense. They point out that animals can detect a wide range of sounds, suggesting that with frequencies ranging from infrasound to ultrasound (frequencies of Hz-kHz) can be perceived by different animals. Some of these sounds can be heard over vast distances. The authors highlight that previous literature reports that "elephants up to 1000 km away from the epicenter detected the breaking of the tsunami on the shores of Sumatra".[110] This has a significant effect on the survival of animals. During the 1994 Sumatran tsunami, elephants detected a noise and then responded to it before the tsunami hit. The sound created by the tsunami would travel at 1260 km/h (the speed of sound under the conditions at the surface of the sea), whereas the waves of the tsunami would travel at 700 km/h, and this, over a distance of 1000 km, gave the elephants a pre-warning of 38.1 minutes. Although many people on the beaches were killed, no elephants were lost or even injured.

Despite the fact that sound is the focus of the research of Garstang and Kelley, they propose that it is not the sound of the earthquake itself that the animals can sense. Rather, what is suggested is that the movement of the Earth's crust causes alterations of the ionic/electron signature of the atmosphere which causes vibration of other objects, and in the abstract of their paper they state:

> The sound heard by animals occurs only when metal or other surfaces (glass) respond to vibrations produced by electric currents induced by distortions of the earth's electric fields caused by the crustal movements. A combination of existing measurement systems

combined with more careful monitoring of animal response could nevertheless be of value, particularly in remote locations.

This relationship between the electronic nature of the effects of the earthquake and the sound produced is known as electrophonics, which can generate a range of sounds with frequencies ranging between 20 Hz and 20 kHz. These are frequencies that can be heard by a range of animals, and therefore it is likely that this is giving the animals the warning that something is amiss, and hence the elephants can run to safety, for example. The speed of the radio waves created would be far faster than either sound waves or the water waves created, and which cause the damage, and hence this theory would account for time of the warning such animals as elephants can detect. It also suggests a way to make a technical solution, i.e., produce something which is a proxy elephant ear.

Whatever the animals are perceiving needs to be turned into a response. Hearing in mammals involves cells which are sensitive to the mechanical action of the sound waves (auditory hair cells), and this is turned into a cell signalling pathway which will lead to neuronal messages being sent to the brain. The intracellular events have been quite extensively studied, and such signalling pathways involve molecules such as guanosine 3'-5'-cyclic monophosphate (cGMP) and the enzymes that make it, i.e. guanylyl cyclases,[111] and then subsequently destroy it i.e. phosphodiesterases, so that the response does not continue forever. Hearing loss in humans continues to be studied, and may involve cell signalling pathways which lead to cell death, or what is often referred to as cellular suicide (apoptosis, otherwise referred to as programmed cell death).[112] Similar cell signalling mechanisms will exist in lower animal species, such as fish.[113] Therefore, understanding how animals use and perceive sound is not difficult to comprehend. Even plants can perceive and respond to mechanical force,[114] and therefore playing music to your plants is not such a mad idea as you might think, and it is often referred to as the Mozart Effect,[115,116] as mentioned earlier.

As well as sound waves in the air, animals use vibrations as an efficient communication system.[117] Amphibians can sense vibrations by putting their heads on the ground, for instance. It is therefore not a surprise if animals can respond to vibrations caused by the movement of the Earth's crust.

In all these cases, whether the animal is sensing sound, magnetism or something we have yet to discover, the animal must have cells which can perceive those physical factors. Once the environmental cue is sensed, the intracellular machinery of the cell will transmit the 'message' received to other cellular components which will initiate transmission of the message to the neuronal system and then to the brain (or what acts as a brain in that organism). A change in action (behaviour) may then ensue, which we can observe.

Why do animals respond to the environmental factors they sense? As discussed earlier, many animals are simply risk adverse. It is better to run than be caught in a catastrophic event. However, animals may also have a sense of knowing that they need to act. As discussed later in this chapter, it is now thought that even lower animals may have some sentience, and not simply acting as a machine.[118] But it must be recognised that behind all these actions is a complex signalling system controlling what the animal, and the cells inside it, do. The cell signalling mechanisms will initiate the response, but also need to be able to be switched off. It is no good if the elephant thinks it needs to run forever. At some point it needs to be able to relax and realise that it is safe. A raft of enzymes is responsible for initiating responses, including kinases (which add phosphate to other molecules), while a suite of enzymes will have the opposite effects, such as phosphatases (which remove phosphate groups). Such signalling will be revisited in the last chapter.

4.7 China and seismic zoos

Previously to the 2009 paper on the Haicheng earthquake (4th February 1975) prediction discussed earlier, the Chinese were collecting data on seismic activity and animal behaviour. The whole topic of using animal observation has been rather big in China, possibly because they have so many earthquakes. A listing suggests that the yearly frequency of earthquakes ranges from eight to 20 in China,[119] whilst the U.S. Geological Survey states: "Both China and Iran are in seismically active areas, have very long historical records, and have had many catastrophic earthquakes."[120] It is thought that the deadliest earthquake recorded anywhere was in Shaanxi, China. It struck on 23rd January 1556, and the death toll was estimated to be 830,000 people.[121]

Faced with such danger, the Chinese have embraced the use of animals as forecasters of future earthquakes. The first experimental station to use biological observations was set up in Hsingtai province in 1968 and then subsequently similar stations were established in Aksu, Sinkiang province, in 1971. Animal behaviours such as snakes coming out of hibernation and staying still on the ground, and rats appearing were reported. Unusual behaviour of larger animals such as horses, cows, pigs and dogs were also reported. Amazingly, using this data, along with geophysical measurements, the State Seismological Bureau of China evacuated the city several hours before the Haicheng earthquake and it was thought that 100,000 lives were saved.[122] Bhargava et al. list several other events where animal behaviour preceded a quake: before the Tangshan Earthquake (magnitude

8.2; 1976); catfish jumping in a pond one day before the before the Edo earthquake (1855). They also cite Tributsch,[123] by saying that examples include:

> dogs barking, nervous cats jumping out of windows, birds scream-
> ing, rats running out of their holes, bees swarming, etc.:

The paper discusses, as earlier, that the tremors are moving as S and P waves and suggests this as an explanation of what the animals are sensing.

How does the Bhargava et al. paper[124] conclude? Despite all the discussion and examples given, the authors say that there are "encouraging results" when it comes to using animal behaviour in earthquake prediction. Hardly a glowing endorsement.

With such scepticism, is the State Seismological Bureau of China still monitoring animals? In 2015 there was a flurry of news articles which shows that the Chinese are seriously looking at animal behaviour. According to the BBC,[125] in Nanjing:

> An ecological park in the city's Yuhuatai district has become one of
> the seismic monitoring sites, with 2,000 chickens, 200 pigs and 2 sq
> km (0.8 sq miles) of fish ponds.

Cameras have been set up around the park and staff working there are encouraged to report any abnormal behaviour. Other news articles reported this similarly. *Phys.org* reports that seven farms are included, and that the animals being monitored also include toads.[126] This was also reported by the *Hindustan Times*[127] as well as other news outlets.[128] The China Earthquake Administration[129] says that part of their remit is "[t]o manage the earthquake monitoring and prediction efforts" and to "[t]o provide guidance to increase awareness about earthquake preparedness", so it looks as though the observations of animals may be an integral part of this. It appears to be quite a major undertaking, and there must be a considerable cost attached to it, so the authorities must think that it is worth it. Perhaps even the smallest chance of predicting the next disaster is worthwhile when the scale of the human cost of doing nothing is taken into account.

The idea of using animals for earthquake prediction has been echoed by Koirala and Acharya. Using survey methods after the 2015 Gorkha Earthquake in Nepal, they suggest that the observation, and more over the training of, animals should be a part of future preparedness in countries such as Nepal and India, as well as other Asian countries.[130] Another, survey-based, investigation was carried out in Japan in 2014.[131] In that study, 1,259 dog owners and 703 cat owners gave a response.

A surprisingly large number of these reported unusual animal behaviour, usually restlessness, in the day before the earthquake: 236 dog owners and 115 cat owners. The authors also looked at milk outputs from farms. It seems incredible, but the farms within 340 km of the epicentre had significantly decreased milk production in the week leading up to the earthquake. Farms farther away were not affected, so acted as a good control group. It was concluded that careful use of such observations in the future could be good for earthquake prediction. This study also suggests that there is something changing which can be sensed up to a week before the earthquake happened.

It seems fitting that the final word on earthquakes and animals – at least for now – should go to Helmut Tributsch, who is now retired from the Free University Berlin, Germany. His 2013 paper in *Animals*[132] starts with the history of human thoughts and animal-based earthquake predictions, and points to a human sacrifice that took place just before a major earthquake on the island of Crete during the Minoan civilisation in 1700 BC. He argues that this was probably not a coincidence. Fascinatingly, he looks at some artwork which may be relevant. In a picture of Bacchus from Pompeii is a dog howling, which Tributsch attributes to the prediction of an earthquake caused by the volcano. The picture also depicts snakes and birds, and he suggests it may have been painted in 62 AD, following another severe earthquake. A second figure shows an "excited dog, and of a fleeing rat", both which were taken prior to the Sungfan-Pingwu earthquake in 1976 (magnitude 7.2).

Tributsch also suggests that some religious ceremonies are earthquake predictions in disguise. He says:

> When wolves penetrated to the Roman Capitol in the year 458 B.C. an expiation ceremony was ordered. People had to leave their houses and to join a procession with decorated cattle.

He goes on to suggest that this was a way to ensure that the population was out in the open, so relatively safe from an earthquake, if it happened. If it never happened: "it was God's will or the result of a successful ceremony aimed at avoiding it."

The Chinese embracing of animal-based prediction is also discussed. Leaflets and even painted fans[133] were distributed which listed animals which need to be observed by the people as they go about their daily business. More than 50 animals were listed, with some being deemed better, i.e., more sensitive, to future earthquakes. Using such information, the Chinese have claimed that they have predicted ten earthquakes, the most famous being the one at Haicheng as discussed earlier. Unfortunately, the Tangshan earthquake on 27th July 1976 (magnitude 7.8) was not publicly anticipated. It caused at least 240,000 deaths and mass destruction. At the time animal-based prediction was thought to not have anticipated this event, but Tributsch appears to have access to internal documents and reports. There is a rather sad story which he relates:

> In the majority of cases the anomalies observed gained importance only after the quake. For example, this happened in the case of a woman from the village of Daodi, in the district of Fengnan. Before she died of her injuries she said that on the evening before the quake the children had insisted that mice were behaving so strangely that an earthquake may be imminent. The woman considered this as nonsense and did not react. Unfortunately the quake arrived and the children did not survive.

Amongst the other animal behaviours listed are dogs barking and removing their puppies from the house, the restlessness of cats, rabbits refusing to eat and trying to leave their pens, chickens flying into the trees and refusing to go into their pens, ducks and geese avoiding water, pigeons avoiding their lofts, and fish swimming in circles and jumping from the water. The list goes on and includes strange behaviour of weasels, mice, frogs, bats, snakes, sable[134] and insects:

> Near the harbour town Tingbo a large oil tanker became literally covered by dragon flies, butterflies, grasshoppers, crickets and cicadas. They could be touched by hand.

It seems unlikely that all such animal behaviours are a coincidence. Furthermore, the unusual animal behaviour started about two days before the earthquake and peaked at eight hours before, so clearly not because of the event itself. When the data were analysed in more detail, the density of anomalous behaviours reported within 5 km of the epicentre (where the magnitude was greater than 7 on the Richter scale) was five times higher than elsewhere. Again, this suggests that there is more than coincidence here. Unfortunately, it appears that the data collected, or at least entered into the report that Tributsch cited,[135] were not used before the Tangshan earthquake, and the result was a significant loss of life, which could potentially have been avoided.

It seems likely, therefore, that the observations of animals prior to an earthquake may be a good predictor for when disaster may strike. There is a wide range of animals listed in the literature here. The reports Tributsch quotes say that 30 animal species were used. Clearly earthquake warning works, regardless of where that prediction might come from. When discussing the pre-event warning in Qinglong in 1976 Tributsch says:

> By 26 July, one day before the disaster, many of Qinglong's 470,000 residents had moved into tents, the rest slept with open doors and windows as a strategy for fast escape. Even though more than 180,000 buildings collapsed, only one person died and that was of a heart attack.[136]

It should be noted that there were many observations made here, including the properties of water, geo-electricity and geo-magnetism. However, animal behaviours were also used, and were clearly an important aspect of the whole prediction system.

4.8 Animals in the prediction of tsunami and volcanic eruptions

With such wide-ranging evidence, albeit some anecdotal, that animals can be used for earthquake prediction, can such observations be used more widely to predict natural disasters? Of course, before we answer that question it should be noted that many of these natural events are connected: earthquakes can cause tsunami; volcanoes are accompanied by seismic activity.

In Enock's paper on indigenous knowledge systems[137] he states:

> Another good example of animal ability to predict disasters could be what happened recently when the Tsunami struck. Despite the loss of 24000 people, wild animals seemed to have escaped the Indian Ocean tsunami, adding weight to the notion that they possess a 'sixth' sense for predicting seasonal quality and impending disasters.

In a paper purporting to be about tsunami prediction, Meenakshi and Juvanna[138] discuss many animal behaviours which have been looked at, but many they chose as examples are about earthquakes and hurricanes, rather than tsunami. However, they do say that birds took to the air, abandoning their nests, and elephants trumpeted and ran to higher ground before a tsunami hit Thailand in 2005. Those brought back before the tsunami were said to have cried. The authors state that:

> Two of the runaway animals brought back from their mountain retreat to the work camp cried throughout the night before the tsunami disaster struck the next morning.

The observation of 'crying' in these elephants is especially interesting when we consider how an elephant's tear production differs from that of the human animal. Our emotional tear production is caused by the parasympathetic nervous system: when we cry, it helps us relax and calms us. In elephants, the exact opposite is true. Elephants drain 'tears' from their temporal glands on the sides of their heads when they are in high levels of *arousal*. These glands are actually modified apocrine sweat glands, activated by the nervous system: they switch on when the elephant needs to be primed and ready for action. Our human tears help us recover from a trauma that's already happened; an elephant's tears get them hyped up for something that's about to occur.[139]

Elephants are very social animals and are known to show distress, and to even help each other. With Asian elephants, it has been reported that one may console another with their trunk, putting it in the other's mouth or by rubbing it on their face.[140] Trumpeting has been seen in Asian and African elephants when another is dying. For example, it was reported that there were "high-frequency vocalizations (trumpets) by an adult female in the presence of a dying calf".[141] Others say that when elephants have been apart and then reunite, they get very excited and are clearly happy to see each other again. Oliver Lamb, who is a geophysicist at the University of North Carolina, said: "They're sort of like dogs who've come back to their owners after being separated for 15 minutes."[142] Noises the elephants made included trumpeting but also stamping their feet, causing their own small seismic tremors. It certainly appears that elephants have sympathy for others

Although elephants appear to show emotions in their facial expressions (this elephant looks very sad), they do not cry 'emotional' tears like humans. Instead, elephant tears indicate that they are 'ready for action', conveying stress, excitement or fear.
(Photo by Amy Elting on Unsplash.)

in the herd and to show emotions. Douglas-Hamilton *et al.* also suggest that trumpeting is thought to be used as a signal of danger, warning others in the vicinity. This correlates with what Meenakshi and Juvanna said about the elephants long before the tsunami hit. Such trumpeting would be heard by other elephants, but it would also be a warning to other animals.

Such vocal warning as those given by elephants will be heard by humans in a wide area, and therefore be a long-range warning. Interestingly, such warnings by animals being used by humans was recently featured in a work of fiction. In Delia Owens's book, *Where the Crawdads Sing*,[143] the main character, Kya, is worried that there is something near:

> She jerked her head around, searching. A footfall in brush. Not a bear, whose large paws squished the debris, but a solid *clunk* in the brambles. Then the crows cawed. Crows can't keep secrets any better than mud; once they see something curious in the forest they have to tell everybody. Those who listen are rewarded: either warned of predators or alerted to food. Kya knew something was up.

Here, the human character is alerted by the birds, but almost certainly other animals will be sensitive to such cues, indicating possible danger, just as they would if elephants had trumpeted, although there are no wild elephants in North Carolina where the story is set.

The prediction of a tsunami strike is certainly important and it appears that animal observations may be useful. In a 2016 conference paper, Virmani and Jain[144] say:

> Tsunami prediction is a social imperative and there is need to carry out research with respect to abnormal marine animal behaviour.

They base this observation on the previous work of others,[145] and the reports of the predictions of a tsunami which hit the Andaman and Nicobar Islands, based on the behaviour of birds and marine life. According to Tiwari and Tiwari, the locals could interpret the sound of the birds as they were calling that "something was wrong".

More recently there have been attempts to mathematically model marine animal behaviour to predict a tsunami.[146] Using observations of sea turtles, it is claimed that the algorithm can give three outputs: alert, pre-alert and no alert. Clearly such work may be useful, if it works robustly, to save countless lives in the future, all based on how animals behave well before disaster strikes.

With earthquakes and tsunami predictions being possible because of animal observations, it is not a great surprise that the same has been suggested to anticipate volcanic eruptions. This, like most of this science, is not particularly new or novel. A *Science* paper of 1925 mentions the uneasiness of animals whilst discussing earthquakes measured at the Hawaiian Volcano Observatory.[147]

Looking at snake bites in the Raung mountains of East Java, Indonesia, with records of 56 patients, it was reported that the incidence of bites reached a peak just before an eruption. It was also noted that the patients at the dr. Koesnadi Hospital, Bondowoso, were well treated and no one died, which is good.[148] This example, although a little extreme, does show that animal behaviour is altered and can be used as a predictor. As people would be presenting with snake bites anyway, as they desire the appropriate treatment, this would not be difficult to monitor and use abnormalities as predictors.

By attaching transmitters to goats on Mount Etna, researchers claim to have been able to predict seven significant eruptions over a two-year period. This is a volcano which regularly erupts, as we discussed earlier. Having said that, on 4th January 2012, for example, the observations of the goats' behaviour gave them six hours of warning before the eruption.[149] The goats are there anyway and equipped with modern technology this seems to be a relatively easily system to implement and has no adverse animal welfare issues. Such a simple approach may be able to save lives on Sicily and elsewhere.

4.9 Climate change and impact on geophysical events

With awareness of global climate change being at a high at the moment, can the preceding information about animals be used in the future and help us to mitigate change which is taking place? Before that can be ascertained, perhaps a more pertinent question is: is there a connection between the changing climate as Earth warms and the occurrence of earthquakes and serious climatic events?[150]

It is thought that some stresses on the surface of the globe can affect the seismic activity occurring. This is likely to be low level activity, rather than major earthquakes, but some conditions such as drought may be important.[151] In Taiwan it has been found that typhoons can trigger what has been called slow earthquakes.[152] Others suggest that the number of earthquakes is reasonably constant, but that occurrences of floods and cyclones is increasing a lot with time and this correlates to the time of the most climate change.[153] Therefore, as the climate changes, perhaps we need to be more vigilant of major environmental events and perhaps watching the animals can help. As I write this there are reports of tornadoes across six states of the USA,[154] with at least 100 people killed. Anecdotally people are saying that this is unprecedented. Whether this is a direct effect of climate change will no doubt be argued about for months, but such events do seem to be becoming more common and more severe.

The opposite is also true: dramatic events can affect the climate. Volcanoes can affect the atmosphere. In a 1913 paper by C.G. Abbot and F.E. Fowle,[155] whilst wondering about streaks seen in the sky, they noted:

> We adopt the view expressed by Dr. Hellmann that the haze in question was due to the eruption of the volcano of Mount Katmai in Alaska in June, 1912.

Furthermore, volcanoes contribute to changes to the climate, often by causing cooling. Ash is thrown into the air, whilst gases, often sulfureous in nature, can lead to sulfate aerosols in the atmosphere. However, the release of carbon dioxide (CO_2) is not particularly significant compared to anthropogenic activity. A statement from the U.S. Geological Survey says: "volcanoes release less than a percent of the carbon dioxide released currently by human activities."[156] What is released from the volcano, however, does have an impact. Research on major eruptions, such as Krakatoa in 1883, have led to a better understanding of the short- and long-term effects, with various estimates of the cooling which occurs.[157] On the other hand, in a 1985 paper looking at the possible cooling after six major eruptions,[158] it has been stated that:

> We conclude that, while volcanic eruptions certainly do not cause a warming of the earth's surface, the evidence that they cause a cooling is not overly impressive either.

Even if cooling is taking place, it is likely to be of a small amount, probably less than 1 °C, and the effect is likely to be short term. On the other hand, when the major government leaders are arguing at COP26 whether the temperature rise can be kept to 1.5 °C rise compared to pre-industrial temperatures, and if possible, below 2 °C, small differences are crucial to overall climatic effects. A volcano making a 1 °C drop in temperature, even if for a short period, may be significant for some regions of the world.

Climate change remains for some an emotive subject but with recent predictions that it is going to get worse and even may be too late to stop,[159] it is a subject which is not going to go away. Whether major geophysical events will be initiated by climate change is yet to be established, but regardless of whether climate change does have an impact, many of the significant events around the world – Mount Vesuvius, the San Andreas Fault etc. – are already 'overdue' and therefore future vigilance is needed, regardless of how anthropogenic activity may alter the future climate.

4.10　Chapter summary

What is clear from the literature is that the relationship between seismic activity, weather events (such as storms and cyclones) and climate change is complex. What is also evident, rather amazingly, is that the observation of animals can be predictors, especially if accompanied by other measurements, such as geo-magnetism. From the literature quoted earlier in this chapter, it can be seen that several authors advocate the observation of animal behaviours. Some suggest using tags on animals so that they may be monitored. Others suggest training animals specifically for this purpose.

With all this in mind Tributsch suggests that:

> Suitable animal observation strategies with high statistical value should be prepared (e.g., variations in egg production, sound in bee hives, permanence of animals in pens, swarm formation in wild species etc.).[160]

What might be considered by many to be quite wacky science may get more credence as more is understood about the correlation of animal behaviour and natural disasters, and what the various animals are sensing. This may be simply perception of vibrations or sound, but it may be that animals are sensing things that we cannot perceive, such as geomagnetic changes. If what animals are perceiving can be robustly determined, it may be possible to build suitable equipment to mimic the animals, which could then be left running with automatic alert systems. The work of Garstang and Kelley, for example, where the radio waves caused by an earthquake, which can then be perceived as sound, could be the basis of a future earthquake detector.[161] Such technological solutions would be so much easier to install and use, and be more robust than relying on a farmer noticing his dog barking, or that the snakes appear "frozen" to the road. The future of this research will no doubt turn to technological solutions, but it in the first instant, it appears that we do have something to learn from the animals.

In the short term, using radiotracers on animals may be a simple solution to the use of animals as sensors. Attaching small trackers to goats, for example, has no welfare issues, and is simple and easy to adopt. Even if one volcanic eruption could be predicted and lives saved it would be worth the effort. If China continues the use of "seismic zoos" it will be interesting to see what the impacts will be. Data from such experiments where a large number, and a wide range, of animals are used will no doubt bring the topic of using animals as sensors for earthquakes to the fore.

In summary, there seems to be a large amount of evidence, albeit much of it anecdotal, which suggests that a wide range of animal species may be useful for the prediction of earthquakes, volcanic eruptions and the onset of a tsunami. These animals are no doubt picking up many cues in the alteration of their behaviours, but it seems to be a subject that should not be ignored. Hopefully, animal observations in the future will save many lives, and understanding the biochemical mechanisms behind how animals are predicting these events may inform the development of future non-animal-based technologies.

Endnotes

1 IRIS/National Science Foundation: www.iris.edu/hq/inclass/fact-sheet/how_often_do_earthquakes_occur (Accessed 03/11/21)

2 Britannica: www.britannica.com/science/Richter-scale (Accessed 23/12/21)

3 Hanks, T.C. and Kanamori, H. (1979) A moment magnitude scale. *Journal of Geophysical Research*, 84(B5), 2348–2350.

4 US Geological Survey: www.usgs.gov/natural-hazards/earthquake-hazards/science/20-largest-earthquakes-world?qt-science_center_objects=0#qt-science_center_objects (Accessed 03/11/21)

5 Richter, C.F. (1935) An instrumental earthquake magnitude scale. *Bulletin of the Seismological Society of America*, 25(1), 1–32.

6 Physics Today: https://physicstoday.scitation.org/do/10.1063/pt.5.031452/full/ (Accessed 03/11/21)

7 Hanks and Kanamori (1979).

8 Das, R., Sharma, M.L., Wason, H.R., Choudhury, D. and Gonzalez, G. (2019) A seismic moment magnitude scale. *Bulletin of the Seismological Society of America*, 109(4), 1542–1555.

9 Britannica: www.britannica.com/science/seismograph (Accessed 03/11/21)

10 Britannica: www.britannica.com/biography/Zhang-Heng (Accessed 03/11/21)

11 Pajak, J. (2005) Signal processing in the "Zhang Heng Seismograph" for remote sensing of impending earthquakes. In *1st International Conference on Sensing Technology November* (pp. 21–23). Palmerston North, New Zealand: http://seismoscope.all shookup.org/remote-sensing-of-earthquakes.pdf (Accessed 20/12/22)

12 New Advent: www.newadvent.org/cathen/11431a.htm (Accessed 03/11/21)

13 In 2021 two earthquakes in rapid succession (a week apart) did hit Scotland. One was magnitude 2.2. However, it was rare enough to be reported in national (UK-wide) new items. This article says that there 200–300 earthquakes across the UK each year, but that 90% are so insignificant that they are not felt. The Guardian: www.theguardian.com/uk-news/2021/nov/20/scotland-second-earthquake-roybridge-highlands-tremor (Accessed 23/12/21)

14 Museo Galileo: https://catalogue.museogalileo.it/biography/FilippoCecchi.html (Accessed 03/11/21)

15 Scientists would refer to this as being in the X, Y and Z planes.

16 National Academy of Sciences: www.nasonline.org/publications/biographical-memoirs/memoir-pdfs/benioff-victor-h.pdf (Accessed 03/11/21)

17 Fracking is a method for extraction of fossil fuels from deep underground by injection of liquids. It is controversial and can cause earth tremors. The BBC: www.bbc.co.uk/news/uk-14432401 (Accessed 23/12/21)

18 NASA: https://moon.nasa.gov/resources/13/apollo-11-seismic-experiment/ (Accessed 03/11/21)

19 Royal Museums Greenwich: www.rmg.co.uk/stories/topics/strange-things-humans-have-left-on-moon (Accessed 23/12/21)

20 A prototype of the lunar buggy was found in a scrapyard and put up for auction. Hemmings: www.hemmings.com/stories/2016/04/11/once-sold-for-scrap-lunar-rover-prototype-now-could-sell-for-150000 (Accessed 03/01/22)

21 Geology.Com: https://geology.com/articles/san-andreas-fault.shtml (Accessed 03/11/21)

22 The Conversation: https://theconversation.com/the-san-andreas-fault-is-about-to-crack-heres-what-will-happen-when-it-does-58975 (Accessed 03/11/21)

23 Britannica: www.britannica.com/event/San-Francisco-earthquake-of-1906 (Accessed 03/11/21)

24 Insider: www.businessinsider.com/big-one-mega-earthquake-what-will-happen-california-san-andreas-2019-8?r=US&IR=T (Accessed 03/11/21)

25 Caltech Science Exchange: https://scienceexchange.caltech.edu/topics/earthquakes/earthquake-early-warning-systems (Accessed 03/11/21)

26 Britannia: www.britannica.com/event/Lisbon-earthquake-of-1755 (Accessed 04/08/22)

27 HiSoUR: www.hisour.com/big-panorama-of-lisbon-national-tile-museum-of-portugal-51110/ (Accessed 04/08/22)

28 Lisbon Portugal Tourism Guide: www.lisbonportugaltourism.com/guide/azulejo-tile-museum.html (Accessed 04/08/22)

29 History: www.history.com/topics/ancient-history/pompeii (Accessed 03/11/21)

30 Sulfur is officially spelt with a 'f' and not 'ph', even in English, rather than just American.

31 Truong, D.H., Eghbal, M.A., Hindmarsh, W., Roth, S.H. and O'Brien, P.J. (2006) Molecular mechanisms of hydrogen sulfide toxicity. *Drug Metabolism Reviews*, 38(4), 733–744.

32 Britannica: www.britannica.com/place/Herculaneum (Accessed 03/11/21)

33 Visit Pompeii: www.visitpompeiivesuvius.com/en/vesuvius (Accessed 03/11/21)

34 The Royal Society: https://royalsocietypublishing.org/doi/pdf/10.1098/rstl.1828.0012 (Accessed 03/11/21)

35 Volcano Discovery: www.volcanodiscovery.com/vesuvius.html (Accessed 15/03/22)

36 Oregon State University: https://volcano.oregonstate.edu/faq/what%E2%80%99s-most-recent-eruption-vesuvius-and-will-it-erupt-again (Accessed 03/11/21)

37 World Population Review: https://worldpopulationreview.com/world-cities/naples-population (Accessed 03/11/21)

38 Hole in the Donut Cultural Travel: https://holeinthedonut.com/2014/06/29/mount-vesuvius-volcano-eruption-italy/ (Accessed 03/11/21)

39 Early Warning of Volcanic eruptions and Earthquakes in the neapolitan area, Campania Region, South Italy: www.unisdr.org/2006/ppew/info-resources/ewc2/upload/downloads/Gasparini2003AbstractEWC2.pdf (Accessed 05/11/21)

40 The Independent: www.independent.co.uk/news/world/europe/mount-vesuvius-emergency-evacuation-eruption-plans-finalised-a7360686.html (Accessed 23/12/21)

41 The Guardian: www.theguardian.com/world/2021/oct/03/canary-islands-volcano-much-more-aggressive-as-new-fissures-erupt (Accessed 03/11/21)

42 Daily Record: www.dailyrecord.co.uk/news/tenerife-tsunami-warning-issued-after-25161421 (Accessed 03/11/21)

43 Sicily Active: www.sicilyactive.com/en/mount-etna-cable-car (Accessed 03/11/21)

44 Volcano Discovery: www.volcanodiscovery.com/etna.html (Accessed 23/12/21)

45 Tsunami Animation: Sumatra, 2004 – YouTube: https://www.youtube.com/watch?v=4yFNOuo_YxI (Accessed 20/12/22)

46 National Ocean Service: https://oceanservice.noaa.gov/facts/tsunami.html (Accessed 03/11/21)

47 History: www.history.com/news/deadliest-tsunami-2004-indian-ocean (Accessed 31/08/22)

48 The Guardian: www.theguardian.com/environment/2022/jun/23/marseille-alexandria-and-istanbul-prepare-for-mediterranean-tsunami (Accessed 04/8/22)

49 Previously known as Madras. The name was officially changed in 1996. Britannica: www.britannica.com/place/Chennai (Accessed 03/11/21)

50 The Times of India: https://timesofindia.indiatimes.com/city/chennai/december-26-2004-when-tsunami-killed-8000-in-tamil-nadu/articleshow/72975364.cms (Accessed 03/11/21)

51 International Tsunami Information Center: http://itic.ioc-unesco.org/index.php?option=com_content&view=category&id=1164&Itemid=1164#:~:text=Tsunami%20Warning%20%2D%20A%20tsunami%20warning,several%20hours%20after%20initial%20arrival (Accessed 03/08/22)

52 US Tsunami Warning System: www.tsunami.gov/ (Accessed 03/08/22)

53 Northwest Pacific Tsunami Advisory Center (NWPTAC): www.jma.go.jp/bosai/map.html#3/17.2/120.8/&contents=pacifictsunami&lang=en- (Accessed 03/08/22)

54 South China Sea Tsunami Advisory Center: http://scstac.oceanguide.org.cn/index.htm (Accessed 03/08/22)

55 ShakeAlert: www.shakealert.org/ (Accessed 03/08/22)

56 Earthquake Warning California: https://earthquake.ca.gov/ (Accessed 03/08/22)

57 Japan Meteorological Agency: www.jma.go.jp/jma/en/Activities/eew.htm (Accessed 03/08/22)

58 The State Council, The People's Republic of China: http://english.www.gov.cn/statecouncil/ministries/202108/09/content_WS611081a2c6d0df57f98de376.html (Accessed 03/08/22)

59 Peng, C., Jiang, P., Ma, Q., Wu, P., Su, J., Zheng, Y. and Yang, J. (2021) Performance evaluation of an earthquake early warning system in the 2019–2020 M 6.0 Changning, Sichuan, China, Seismic Sequence. *Frontiers in Earth Science*, 9, 699941.

60 OSU-CT, Osservatorio Sismico Urbano: www.ct.ingv.it/osuct/index.php/en/news-eng/49-earthquake-early-warning-eew (Accessed 03/08/22)

61 USGS: www.usgs.gov/programs/VHP/national-volcano-early-warning-system-monitoring-volcanoes-according-their-threat (Accessed 03/08/22)

62 Japan Meteorological Agency: www.data.jma.go.jp/multi/volcano/index.html?lang=en (Accessed 03/08/22)

63 American Geophysical Union: https://blogs.agu.org/geospace/2018/11/28/new-automated-volcano-warning-system-forecasts-imminent-eruptions/ (Accessed 03/08/22)

64 Meterologix: https://meteologix.com/cn/satellite/china/volcano-alert-10min/20220728-1810z.html (Accessed 03/08/22)

65 Quammen, D. (1985) Animals and earthquakes. *This World, San Francisco Chronicle*, April 21, 15–16.

66 US Geological Survey: www.usgs.gov/natural-hazards/earthquake-hazards/science/animals-earthquake-prediction?qt-science_center_objects=0#qt-science_center_objects (Accessed 11/11/21)

67 Tributsch, H. (1982) *When the Snakes Awake: Animals and Earthquake Prediction*. MIT Press, Cambridge, MA, 248 pp.

68 "Uncle Offa" (1991) *Natural Weather Wisdom* (p. 133). The Self Publishing Association Ltd., Worcs, UK.

69 CBC Kids: www.cbc.ca/kids/articles/8-animals-that-give-the-weather-report (Accessed 30/06/22)

70 Wikelski, M., Mueller, U., Scocco, P., Catorci, A., Desinov, L.V., Belyaev, M.Y., Keim, D., Pohlmeier, W., Fechteler, G. and Martin Mai, P. (2020) Potential short-term earthquake forecasting by farm animal monitoring. *Ethology*, 126(9), 931–941.

71 Sciencing: https://sciencing.com/seagull-behaviors-earthquakes-changes-weather-23098.html (Accessed 12/11/21)

72 Public Broadcasting Service: www.pbs.org/wnet/nature/can-animals-predict-disaster-tall-tales-or-true/131/ (Accessed 05/11/21)

73 Public Broadcasting Service: www.pbs.org/wnet/nature/can-animals-predict-disaster-tall-tales-or-true/131/ (Accessed 05/11/21)

74 From California Geology: www.johnmartin.com/earthquakes/eqpapers/00000072.htm (Accessed 05/11/21)

75 Grant, R.A. and Halliday, T. (2010) Predicting the unpredictable; evidence of pre-seismic anticipatory behaviour in the common toad. *Journal of Zoology*, 281(4), 263–271.

76 Yosef, R. (1997) Reactions of Grey Herons (*Ardea cinerea*) to seismic tremors. *Journal für Ornithologie*, 138(4), 543–546. (Note this paper is in English.)

77 Circadian rhythms are those that follow a 24-hour pattern, such as sleep and waking.

78 Yokoi, S., Ikeya, M., Yagi, T. and Nagai, K. (2003) Mouse circadian rhythm before the Kobe earthquake in 1995. *Bioelectromagnetics*, 24(4), 289–291.

79 Li, Y., Liu, Y., Jiang, Z., Guan, J., Yi, G., Cheng, S., Yang, B., Fu, T. and Wang, Z., 2009. Behavioral change related to Wenchuan devastating earthquake in mice. *Bioelectromagnetics: Journal of the Bioelectromagnetics Society, the Society for Physical Regulation in Biology and Medicine, the European Bioelectromagnetics Association*, 30(8), 613–620.

80 Chen, L.L., Hu, X., Zheng, J., Zhang, H.H., Kong, W., Yang, W.H., Zeng, T.S., Zhang, J.Y. and Yue, L. (2010) Increases in energy intake, insulin resistance and stress in rats before Wenchuan earthquake far from the epicenter. *Experimental Biology and Medicine*, 235(10), 1216–1223.

81 Lighton, J.R. and Duncan, F.D. (2005) Shaken, not stirred: A serendipitous study of ants and earthquakes. *Journal of Experimental Biology*, 208(16), 3103–3107.

82 Bhargava, N., Katiyar, V.K., Sharma, M.L. and Pradhan, P. (2009) Earthquake prediction through animal behavior: A review. *Indian Journal of Biomechanics*, 78, 159–165.

83 Smith, S. (2008) Rumor and the Sichuan Earthquake. *The China Beat Blog Archive 2008–2012*, 240: http://digitalcommons.unl.edu/chinabeatarchive/240 (Accessed 11/11/21)

84 Devaney, A.J. and Oristaglio, M.L. (1986) A plane-wave decomposition for elastic wave fields applied to the separation of P-waves and S-waves in vector seismic data. *Geophysics*, 51(2), 419–423.

85 The BBC: www.bbc.co.uk/bitesize/guides/zswkjty/revision/3 (Accessed 11/11/21)

86 Kirschvink, J.L. (2000) Earthquake prediction by animals: Evolution and sensory perception. *Bulletin of the Seismological Society of America*, 90(2), 312–323.

87 Kalmijn, A.J. (1974) The detection of electric fields from inanimate and animate sources other than electric organs. In A. Fessard (ed.), *Handbook of Sensory Physiology* (Vol. 9, pp. 147–200). Springer-Verlag, Berlin, New York.

88 Blakemore, R.P. (1975) Magnetotactic bacteria. *Science*, 190, 377–379.

89 Walker, M.M., Kirschvink, J.L., Chang, S.-B.R. and Dizon, A.E. (1984) A candidate magnetic sense organ in the Yellowfin Tuna *Thunnus albacares*. *Science*, 224, 751–753.

90 Walcott, C. (1978) Anomalies in the earth's magnetic field increase the scatter of pigeon's vanishing bearings. In K. Schmidt-Koenig and W.T. Keeton (eds.), *Animal Migration, Navigation and Homing* (pp. 143–151). Springer-Verlag, Berlin.

91 Gould, J.L., Kirschvink, J.L. and Deffeyes, K.S. (1978) Bees have magnetic remanence. *Science*, 201, 1026–1028.

92 Karki, N., Vergish, S. and Zoltowski, B.D. (2021) Cryptochromes: Photochemical and structural insight into magnetoreception. *Protein Science*. (*in press*)

93 Fleischmann, P.N., Grob, R. and Rössler, W. (2020) Magnetoreception in Hymenoptera: Importance for navigation. *Animal Cognition*, 1–11.

94 Walker, M.M., Dennis, T.E. and Kirschvink, J.L. (2002) The magnetic sense and its use in long-distance navigation by animals. *Current Opinion in Neurobiology*, 12(6), 735–744.

95 Odum, H.T. (1948) The bird navigation controversy. *The Auk*, 65, 584–597.

96 Named after Gustave-Gaspard Coriolis, the Coriolis Force is created when a body rotates.

97 Keeton, W.T. (1971) Magnets interfere with pigeon homing. *Proceedings of the National Academy of Sciences*, 68, 102–106.

98 Harada, Y. (2002) Experimental analysis of behavior of homing pigeons as a result of functional disorders of their lagena. *Acta oto-laryngologica*, 122, 132–137.

99 Harada, Y. (2008) The relation between the migration function of birds and fishes and their lagenal function. *Acta oto-laryngologica*, 128, 432–439.

100 Massa, B., Ioalè, S.B.P., Lo Valvo, M. and Papi, F. (1991) Homing of Cory's shearwaters (*Calonectris diomedea*) carrying magnets. *Italian Journal of Zoology*, 58, 245–247.

101 Bonadonna, F., Chamaillé-Jammes, S., Pinaud, D. and Weimerskirch, H. (2003) Magnetic cues: Are they important in Black-browed Albatross *Diomedea melanophris* orientation? *Ibis*, 145, 152–155.

102 Mouritsen, H., Huyvaert, K.P., Frost, B.J. and Anderson, D.J. (2003) Waved albatrosses can navigate with strong magnets attached to their head. *Journal of Experimental Biology*, 206, 4155–4166.

103 Papi, F., Luschi, P., Akesson, S., Capogrossi, S. and Hays, G.C. (2000) Open-sea migration of magnetically disturbed sea turtles. *Journal of Experimental Biology*, 203, 3435–3443.

104 Tian, L., Luo, Y., Zhan, A., Ren, J., Qin, H. and Pan, Y. (2022) Hypomagnetic field induces the production of reactive oxygen species and cognitive deficits in mice hippocampus. *International Journal of Molecular Sciences*, 23, 3622.

105 The term respiratory burst is a misnomer as it has nothing to do with respiration. Oxygen is used but converted to a radical called superoxide (O_2^-), which can then be used to make H_2O_2 and hypochlorous acid. The enzyme involved is called NADPH oxidase. Lack of this activity leads chronic granulomatous disease (CGD), where patients struggle to control pathogen attacks, especially in their lungs.

106 Raipuria, R.K., Kataria, S., Watts, A. and Jain, M. (2021) Magneto-priming promotes nitric oxide via nitric oxide synthase to ameliorate the UV-B stress during germination of soybean seedlings. *Journal of Photochemistry and Photobiology B: Biology*, 220, 112211.

107 A molecule is a radical if it has an unpaired electron in its outer orbital. This makes such molecules relatively reactive. Interestingly, molecular oxygen (O_2) is a di-radical and yet is stable in air.

108 Journals often run Special Issues, or Topics, in areas which academics think are hot at the moment. A Topic has just opened entitled *Magnetobiology and Magnetomedicine*, with a closing date of August 2023, so new papers should be published in this area over the next year: www.mdpi.com/topics/Magnetobiology_Magnetomedicine (Accessed 12/08/22)

109 Garstang, M. and Kelley, M.C. (2017) Understanding animal detection of precursor earthquake sounds. *Animals*, 7, 66.

110 Garstang, M. (2009) Precursor tsunami signals detected by elephants. *Open Conservation Biology Journal*, 3, 3.

111 Marchetta, P., Rüttiger, L., Hobbs, A.J., Singer, W. and Knipper, M. (2022) The role of cGMP signalling in auditory processing in health and disease. *British Journal of Pharmacology*, 179, 2378–2393.

112 Wu, J., Ye, J., Kong, W., Zhang, S. and Zheng, Y. (2020) Programmed cell death pathways in hearing loss: A review of apoptosis, autophagy and programmed necrosis. *Cell Proliferation*, 53, e12915.

113 Higgs, D.M. (2020) Functional review of hearing in zebrafish. In *Behavioral and Neural Genetics of Zebrafish* (pp. 73–91). Academic Press, Cambridge, MA (imprint of Elsevier).

114 Sparke, M.A. and Wünsche, J.N. (2020) Mechanosensing of plants. *Horticultural Reviews*, 47, 43–83.

115 Exbrayat, J.M. and Brun, C. (2019) Some effects of sound and music on organisms and cells: A review. *Annual Research & Review in Biology*, 1–12.

116 Jenkins, J.S. (2001) The Mozart effect. *Journal of the Royal Society of Medicine*, 94, 170–172.

117 Hill, P.S. (2001) Vibration and animal communication: A review. *American Zoologist*, 41, 1135–1142.

118 Lambert, H., Elwin, A. and D'Cruze, N. (2021) Wouldn't hurt a fly? A review of insect cognition and sentience in relation to their use as food and feed. *Applied Animal Behaviour Science*, 243, 105432.

119 Statista: www.statista.com/statistics/224557/number-of-earthquakes-in-china/ (Accessed 12/11/21) Note: there is no definition of an earthquake given here.

120 US Geological Survey: www.usgs.gov/faqs/which-country-has-most-earthquakes?qt-news_science_products=0#qt-news_science_products (Accessed 12/11/21)

121 History: www.history.com/this-day-in-history/deadliest-earthquake-in-history-rocks-china (Accessed 23/12/21)

122 Bhargava *et al.* (2009).

123 Tributsch (1982).

124 Bhargava *et al.* (2009).

125 The BBC: www.bbc.co.uk/news/blogs-news-from-elsewhere-33362592 (Accessed 11/11/21)

126 Phys Org: https://phys.org/news/2015-07-china-animals-earthquakes.html (Accessed 11/11/21)

127 Hindustan Times: www.hindustantimes.com/world/animals-being-used-to-predict-earthquakes-in-china-report/story-Oq1dWw4bD3shOtInihvrGJ.html (Accessed 11/11/21)

128 ABC News: www.abc.net.au/news/2015-07-06/china-using-animals-to-predict-earthquakes-reports-say/6599366 (Accessed 11/11/21)

129 The State Council, The People's Republic of China: http://english.www.gov.cn/state_council/2014/10/01/content_281474991089800.htm (Accessed 11/11/21)

130 Koirala, J. and Acharya, S. (2015) Can birds and animals predict disaster? A case study from Nepal earthquake 2015. *A Case Study from Nepal Earthquake*: https://papers.ssrn.com/sol3/papers.cfm?abstract_id=3792633 (Accessed 11/11/21)

131 Yamauchi, H., Uchiyama, H., Ohtani, N. and Ohta, M. (2014) Unusual animal behavior preceding the 2011 earthquake off the Pacific coast of Tohoku, Japan: A way to predict the approach of large earthquakes. *Animals*, 4, 131–145.

132 Tributsch, H. (2013) Bio-mimetics of disaster anticipation – learning experience and key-challenges. *Animals*, 3, 274–299.

133 Tributsch (2013) shows pictures of these, Figures 3 & 4.

134 Sable (*Martes zibellina*) is a species of marten.

135 Mei, S., Hu, C., Zhu, C., Ma, J., Zhang, Z. and Yang, M. (1982) *The Tangshan Earthquake of 1976*. Seismological Press, Beijing, China. [Chapter XII in particular]

136 Tributsch (2013).

137 Enock, C.M. (2013) Indigenous knowledge systems and modern weather forecasting: Exploring the linkages. *Journal of Agriculture and Sustainability*, 2(2).

138 Meenakshi, A.N. and Juvanna, B.I. (2013) Tsunami detection system using unusual animal behavior- A specified approach. *Earth Science*, 2(1), 9–13.

139 Garner (2018) Elephants don't have tear ducts – so why are they always crying? Why Animals Do The Thing blog: www.whyanimalsdothething.com/elephants-dont-cry (Accessed 27/09/22)

140 Plotnik, J.M. and de Waal, F.B. (2014) Asian elephants (*Elephas maximus*) reassure others in distress. *PeerJ*, 2, e278.

141 Douglas-Hamilton, I., Bhalla, S., Wittemyer, G. and Vollrath, F. (2006) Behavioural reactions of elephants towards a dying and deceased matriarch. *Applied Animal Behaviour Science*, 100, 87–102.

142 Atlas Obscura: www.atlasobscura.com/articles/elephants-and-seismometers (Accessed 04/08/22). More of Lamb's work can be found here: Lamb, O.D., Shore, M.J., Lees, J.M., Lee, S.J. and Hensman, S.M. (2021) Assessing raspberry shake and boom sensors for recording African elephant acoustic vocalizations. *Frontiers in Conservation Science*, 10. However, unfortunately the authors also had to publish a corrigendum: www.frontiersin.org/articles/10.3389/fcosc.2021.664303/full (Accessed 04/08/22)

143 Owens, D. (2019) *Where the Crawdads Sing*. Corsair, London. ISBN: 9781472154668. Also now a major film: www.imdb.com/title/tt9411972/ (Accessed 31/08/22)

144 Virmani, D. and Jain, N. (2016) Intelligent information retrieval for Tsunami detection using wireless sensor nodes. In *2016 International Conference on Advances in Computing, Communications and Informatics (ICACCI)* (pp. 1103–1109). IEEE, London.

145 Tiwari, R. and Tiwari, S. (2011) Animals: A natural messenger for disasters. *Journal of Natural Products*, 4, 3–4. They also quote this article: Central Chronicle (2005): Traditional alert 'saved Andaman tribes'.

146 Jain, N., Virmani, D. and Abraham, A. (2019) Overlap function based fuzzified aquatic behaviour information extracted tsunami prediction model. *International Journal of Distributed Systems and Technologies (IJDST)*, 10(1), 56–81.

147 Finch, R.H. (1925) An earthquake prediction at Hawaiian Volcano Observatory. *Science*, 61(1567), 42–43.

148 Kurniawan, N., Kadafi, A.M., Kurnianto, A.S., Ardiansyah, F. and Maharani, T. (2017) Understanding snake bite cases pattern related to volcano-seismic activity: An evidence in Bondowoso, Indonesia. *Biotropika: Journal of Tropical Biology*, 5(3), 102–109.

149 Icarus Global Monitoring with Animals: www.icarus.mpg.de/28810/animals-warning-sensors#:~:text=more%5D-,Animals%20can%20sense%20natural%20dis asters%3A%20Goats%20on%20Mount%20Etna%20in,of%20imminent%20erup tions%20in%20future.&text=Animal%20Behavior%2F%20MaxCine-,Ani mals%20can%20sense%20natural%20disasters%3A%20Goats%20on%20 Mount%20Etna%20in,become%20anxious%20before%20major%20eruptions (Accessed 12/11/21)

150 NASA: https://climate.nasa.gov/news/2926/can-climate-affect-earthquakes-or-are-the-connections-shaky/ (Accessed 04/11/21)

151 NASA: https://climate.nasa.gov/news/2926/can-climate-affect-earthquakes-or-are-the-connections-shaky/ (Accessed 04/11/21)

152 Liu, C., Linde, A. and Sacks, I. (2009) Slow earthquakes triggered by typhoons. *Nature*, 459, 833–836.

153 Peduzzi, P. (2005) Is climate change increasing the frequency of hazardous events? *Environment and Poverty Times*, (3), 7.

154 The BBC: www.bbc.co.uk/news/world-us-canada-59623970 (Accessed 12/12/21)

155 Abbot, C.G. and Fowle, F.E. (1913) Volcanoes and climate. *Smithsonian Miscellaneous Collections*: https://platforma.bk.pan.pl/en/search_results/475896 (Accessed 20/12/22)

156 US Geological Survey: www.usgs.gov/natural-hazards/volcano-hazards/volca noes-can-affect-climate (Accessed 04/11/21)

157 Schaller, N., Griesser, T., Fischer, A., Stickler, A. and Brönnimann, S. (2009) Climate effects of the 1883 Krakatoa eruption: Historical and present perspectives. *Vierteljahrsschrift der Naturforschenden Gesellschaft in Zürich*, 154, 31–40.

158 Angell, J.K. and Korshover, J. (1985) Surface temperature changes following the six major volcanic episodes between 1780 and 1980. *Journal of Climate and Applied Meteorology*, 24(9), 937–951.

159 McGuire, B. (2022) *Hothouse Earth: An Inhabitant's Guide*. Icon Books, London. ISBN: 978-1785789205

160 Tributsch (2013).

161 Garstang and Kelley (2017).

5

Solar and lunar activity

5.1 Introductory story

The herdsman was feeling cold, but he was nearly home. He knew that a hot dinner of reindeer meat would be ready for him. It had been a long trip. The herd had moved far further west than they should have, and his team had had to work hard to find all the animals and round them up. Using the sledges and dogs, they had then herded nearly 200 reindeer back to the village. And now there was just the lake between him and a hot meal and a warm bed. He was also looking forward to seeing his wife and children, having been away for so long.

He knew that the whole surface of the lake would be frozen solid and that the ice would be deep. It had started to cover at least two moons before, and there was no chance that the ice had thinned since. He confidently headed out from the bank, charting a course straight across the frozen water. In another hour he would be home.

The herdsman abruptly stopped. The green glow started from the north and the sky started to swirl. His ancestors were stirring again. This was not a good sign. Had they been upset? Had he or the villagers been disrespectful in some way? Or was this a warning?

As he watched, the green changed to a yellow and then a slightly pink glow appeared. He was mesmerized, even though he had seen the night light up hundreds of times before. There was never any warning. He never knew when the glow would come, and how long it would last.

The dog sitting next to him looked up too. As they both watched, a face of a monkey appeared in the sky. It did not look happy. Something had upset the ape and slowly it morphed into the face of a witch. The herdsman started to shiver in fright. It was said that the strange lights were his ancestors having a game, but this did not look like a happy contest. He was sure it was trying to tell him something. However, as he watched, the glow disappeared as quickly it had arrived, and the herdsman was left looking up at the stars, innocently looking down at him, as though there was nothing wrong.

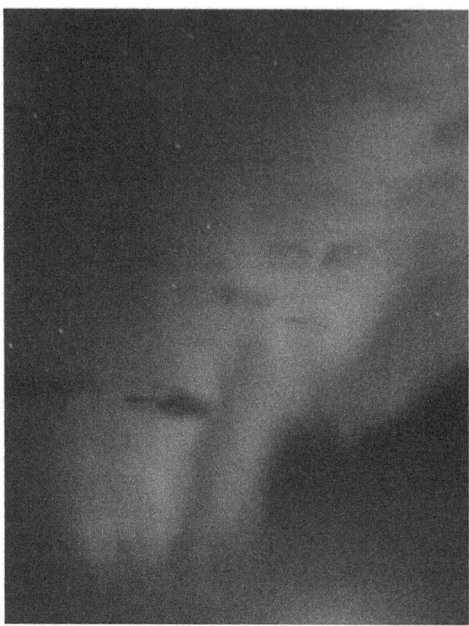

The Northern Lights at Äkäslompolo in 2021. Note the "monkey face" towards the top right.
(Photo the author's, using a Canon digital camera. It has been enhanced for clarity and made B/W, which is how some animals may see it.)

Now sure what he had to do, the herdsman struck out on a new course along the edge of the lake. There was something untoward ahead, and he would take the long route home, even though it would delay that well-anticipated meal and hugs from his kids. The lights rarely lied and if it took him twice as long to get home, so be it. It was better to arrive safely than to fall through the ice to certain death.

5.2 Solar and lunar events – an introduction

This book started by describing the behaviour of horses prior to a solar eclipse. So, do animals have any ability to either respond to or predict the activity of the moon and stars? Or was this – what I perceived as – odd behaviour either my imagination or a coincidence?

People have been staring upwards forever. It is hard to look at a clear night and not ponder what is out there, or indeed, if there are any other life forms there, perhaps looking at us. Certainly, that debate appears to continue and is not likely to abate any time soon.[1]

An eclipse is a bizarre event if you do not know what is happening.[2] Imagine if you lived thousands of years ago. You were out planting your crops on a nice sunny day, and then in the middle of that day the sun quite suddenly disappears. You may be puzzled, and perhaps frightened. Perhaps you have upset the gods and they have taken the sun away? The event is so uncommon it must have come as a surprise. Although they probably do not have the concept of a god, this might be how a non-human animal might feel when a solar eclipse occurs. They are likely to be very confused and frightened. Would they therefore initiate a self-preservation response? Or perhaps they knew it was about to happen?

Solar eclipses are not the only events that animals may be confused about. Lunar eclipses, when the moon moves into the Earth's shadow and disappears, must also be odd if a nocturnal animal is out and about in the moonlight. Records of such lunar events have been kept for centuries,[3] and no doubt mystified humans for a long time.

The Northern Lights (The Arora Borealis: *Aurora Australis* (Southern Lights) in the southern hemisphere) are amazing to see, but they come and go with what appears to be randomness. Of course, humans now know what is occurring, and can even predict when it is likely to happen, as discussed later, but it must be odd to an animal. Animals may not be able to see the colours of the aurora that we can see, or they may be able to see colours outside the range that we are sensitive to, so the Arora Borealis may look very different to them. Even if they only have black and white vision, they are likely to be able to see a strange glow in the sky in the middle of the night. It must be very strange for the night to become lighter, especially as it moves and creates swirling patterns.

The questions which would be pertinent here are two-fold. Do animals respond to events in the sky? Do animals predict that these events are likely to happen in the near future? And perhaps a third question could be: do animals care about such events at all? For me, the question on my mine was: did the horses know that a solar eclipse was imminent?

Here, reports in the literature and anecdotal evidence of whether animals can be of any use in predicting celestial events will be reviewed.

5.3 Instrumentation which has been developed to observe the skies

To look out to the stars and planets, it is important to have a good telescope. The first instrument was said to have been invented by Galileo Galilei (1564–1642). Although he was probably *not* the first, he was certainly a person who made one useable. During Galileo's life, a Dutch spectacle maker, Hans Lipperhey (or Lipperhey: circa 1570–1619), produced a tele-scope (in 1608),[4] although there were some thoughts that he might have copied another local optic maker called Zacharias Jansen.[5] Lipperhey may have watched children playing with lens and got the idea of putting them together into a telescope. His instrument was able to magnify objects three-fold. At more or less the same time as Lipperhey applied for his pat-ent, another person, also Dutch, had applied for a patent for a very similar instrument. This was Jacob Metius. It seemed as though the time was ripe for the invention of this instrument. However, the telescope with a lens at each end is often attributed to Galileo and is often referred to as such, even though Galileo was thought to have known about the Lipperhey instru-ment. Galileo's was much better, having a magnification of about 20-fold. He, himself, made momentous observations of the moon and stars, many of which he captured in amazingly detailed drawings. He was the first to identify craters and mountains on the moon, and also observed the moons of Jupiter and the rings of Saturn. However, famously his ideas fell afoul of the Catholic Church, and he was placed under house arrest for nine years. His idea of the sun being the centre of the solar system was first sug-gested by the Polish mathematician Nicolaus Copernicus (1473–1543)[6] in his work entitled *De Revolutionibus Orbium Coelestium* (*On the Revolutions of the Celestial Spheres*), published on his death in 1543. Although Galileo ded-icated his work (*Dialogue Concerning the Two Chief World Systems, Ptolemaic and Copernican*) to Pope Urban VIII, Galileo was declared as a heretic and had to appear at an inquisition in Rome in 1633. Even though he was found guilty and sentenced to house arrest, it was reported he had said words along the lines of "*Eppur si muove*" ("And yet it moves") in front of the inquisitors as he was being prosecuted, indicating that he still believed that the Earth was not static in the universe. However, this is almost cer-tainly a myth[7] and will never be able to be substantiated! His days under house arrest are extremely well captured in the book *Galileo's Daughter*, by Sobel,[8] which is based on correspondence between Galileo and his daugh-ter, who lived in a convent as the nun Suor Maria Celeste (Virginia Galilei (1600–1634)).

The next great leaps forward in telescope technology were probably by the German Johannes Kepler (1571–1630)[9] and the Englishman Sir Isaac Newton (1643–1727),[10] the latter who invented what was to be referred to as the Newtonian Telescope. This had an open end, but the other end was not a lens but a mirror. This reflected the image back to a second mirror which sent the light into a lens which was on the side of the main tube. This could effectively double the length of the telescope and improve the light capture.

Both the Galileo and Newtonian telescopes are still popular today, but for more serious star gazing much bigger and more elaborate telescopes have been invented. Famously, Sir William Herschel had a 40-foot reflecting telescope built in his garden,[11] which became something of a tourist attraction. This was described in one of his papers in 1795, published in the *Philosophical Transactions of the Royal Society of London*.[12] The paper starts:

> The uncommon size of my forty-feet reflecting telescope will render a description of it not unacceptable to lovers of astronomy.

He was obviously very proud of it. Herschel was an extremely serious astronomer and hugely successful. He was the first person to see Uranus, which he originally named *Georgium Sidus* in honour of the British King. It should be noted that he was also extremely ably assisted by his sister, Caroline,[13] and together they would spend many hours collecting data.

Harlow Shapley, as discussed in Chapter 2, was a renowned astronomer, but has been discussed here because of his observations of the speed of ants. He was at Mount Wilson[14] at the time. That telescope facility was instigated in 1904 by George Ellery Hale[15] (1868–1938). In 1908 a 152-cm

Herschel's Grand Forty Feet Reflecting Telescope. Note the building, along with two people, at the bottom left which gives a sense of scale.
(Sourced from Picryl. This image has been made black and white.)

(60-inch) reflector telescope was installed. At the time it was the largest in the world, and it was this instrument to which Shapley had access. In 1917 the 100-inch (2.5m) Hooker telescope was installed at Mount Wilson, and this became the largest until 1949, and was used by Edwin Hubble.[16] Hubble changed the way people thought about the universe, discovering galaxies outside of our own Milky Way – hence why the Hubble Space Telescope is named after him.

However, such telescope technology could not last as the only option. Mirrors were generally made from alloy of copper and tin called *speculum*, and it became more and more difficult to make large versions. Many gigantic telescopes were produced, but in the 1930s the technology took a major shift. Karl Guthe Jansky, an engineer at Bell Telephone Laboratories, built a short-wave detector. This instrument measured 30 metres across and 6 metres tall, and earned the nickname of the "merry-go-round".[17] As well as "hearing" a hiss from thunderstorms Jansky found a signal which he suspected emanated from space. Building on this work, an amateur enthusiast called Grote Reber in 1937 made a parabolic telescope (otherwise known as a dish telescope) in his garden in Wheaton, Illinois, much as Heschel had done. This instrument was 9 metres across. The radio telescope was born.

Twenty years later, at Jodrell Bank, Cheshire, UK, Sir Bernard Lovell built a dish radio telescope which was 76 metres across. In total the telescope is 3,200 tonnes, and to re-paint it with three coats needs 5,300 litres of paint.[18] It is a far cry from the early telescopes of Hans Lipperhey, and if one wants to know what the Jodrell Bank telescope is doing, now there is even a webcam to watch it.[19]

However, telescopes did not remain grounded, but have been sent into space to send the images back. This avoids the interference from the atmosphere, a problem which was on the mind of many astronomers through history when they wished to site their new instruments.[20] It was also on our minds as we waited for the eclipse in 1999, as clouds would ruin the day. Clouds may have also afforded Shapley time to study his ants, so there is often a positive outcome too.

The Hubble Space Telescope was launched in 1990 as a joint venture by the National Aeronautics and Space Administration (NASA) and the European Space Agency (ESA). It has returned to Earth some truly amazing images,[21] and one can even see some of the data live.[22] Eventually Hubble will be replaced by the James Webb Space Telescope. As I write this (18th November 2021) it is 29 days and 18 hours from launch,[23] so by the time you read this it will be in orbit.[24,25] Now in August 2022, the James Webb Space Telescope is already returning to Earth truly amazing pictures, with light emanating from the deeper reaches of the universe.[26]

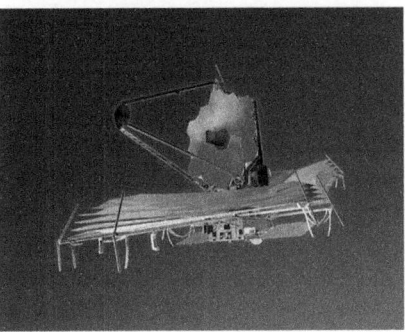

The Hubble Space Telescope, 2002, and NASA's artwork of the James Webb Space Telescope.
(From NASA sourced through Picryl. Both images have been made black and white.)

Other notable telescopes include the Herschel Space Observatory built by ESA.[27] It was the largest ever infrared telescope launched into space and operated from 2009 to 2013. An extremely large telescope is the W. M. Keck Observatory[28] near the summit of Mauna Kea (often given as one word, Maunakea, which is probably more traditional in the local culture[29]) on the Island of Hawaii, in the Hawaiian archipelago. This is a twin-telescope.

The world's largest telescope is the Five-hundred-meter Aperture Spherical Telescope (FAST),[30] which is in a karst[31] depression in Guizho, a mountainous province of southwest China. This telescope is said to have a surface area which is equivalent to 30 football pitches. The largest optical telescope is the Gran Telescopio Canarias[32] (GTC) (La Palma, Canary Islands). This has an aperture of 10.4 metres.[33] Construction started in 2002, took seven years and cost €130 million.

With telescopes getting ever more complicated and expensive, is there anything that observing the animal world can tell us? It seems unlikely, but I still think back to the horses at the 1999 eclipse.

5.4   Animals and the events in the sky

To see if the activity of white-tailed deer (*Odocoileus virginianus*) were affected by lunar patterns, rather just following dawn and dusk, animals were equipped with GPS collars.[34] In the abstract of their paper, the authors state:

> on those days furthest from the full or new moon, deer were less likely to be active during moonrise and moonset periods, and more likely to be active during moon overhead and moon underfoot periods.

This is, according to others, widely known and used as an aid in hunting,[35] and was the subject of a book by Jeff Murray entitled *Moon Struck; Hunting Strategies That Revolve Around the Moon.*[36] Murray also wrote further on this topic, with *Moon-Phase Deer Hunting*, to help fellow hunters maximise their chances of success.[37] Clearly there was enough of a relationship between the lunar cycles and behaviours of deer to be of pragmatic use.

In 2018 Navas González *et al.* asked, "Can donkey behavior and cognition be used to trace back, explain, or forecast moon cycle and weather events?"[38] In the 'simple summary' of their paper they have a very strong claim:

> Conclusively, donkeys can be used as an environment informative sensitive tool and may therefore, predict and register slight human-unappreciable climatic variations to which they may behaviorally adapt beforehand.

That would go some way to explain my horse story at the eclipse, as donkeys and horses are both members of the Equidae family of species. So, what did these authors do to be so certain? Using 300 donkeys (Andalusian donkey breed) they observed 14 environmental factors whilst at the same time measuring the mood and responses of the animals, along with their learning capacity. The weather parameters recorded included the temperature, relative humidity, windspeed, sunlight hours and barometric pressure. They also recorded rainfall per day but also interestingly predicted rainfall. Lunar data were obtained from the US Naval Observatory. By looking at conditions at birth of the animals along with current conditions at the time of the experiments, the authors concluded that along with other environmental conditions that moon cycle phases were amongst stress factors that could affect the animals, and have long lasting effects. They pointed to two effects that the moon might be having: difference in light levels and effects on gravity. After all, the tides are determined by the lunar gravitational pull.[39]

Interestingly, in 2018 Navas González *et al.*[40] also suggested that their data was one of the first to give "scientifically proof" to the myth that the behaviour of donkeys could be used as a predictor of bad weather, in particular oncoming rain.

It is also interesting that the study on donkeys recorded the lunar phase at the birth of the animals. In cows this was also found to have an effect.[41] In this study, the authors observed domestic Holstein cows, over a three-year period at a private farm in Hokkaido, Japan. They compared the records of 428 spontaneous, full-term deliveries to the lunar cycles at the time. The authors then state:

> Spontaneous birth frequency increased uniformly from the new moon to the full moon phase and decreased until the waning crescent phase.

This seems amazing. However, the topic was reviewed in 2006 by Michał Zimecki,[42] who suggests the lunar effects are seen in insects, birds, rats, mice, sheep and fish. He admits that the exact action here is not known but could be down to changes in electromagnetic radiation or the moon's gravity. Others agree, and even though the reproduction of fish is affected by the moon the exact molecular mechanism remains somewhat elusive, although hormonal changes do seem to be involved.[43]

Spawning events in amphibians were found to be more frequent around the full moon rather than at the time of the new moon.[44] Whilst toads and frogs were thought to be responding to lunar light, or perhaps have some internal synchronised lunar cycle, in newts it was suggested that they may be responding to changes in gravity or geomagnetic forces. Such alignment with the lunar cycle was thought to enable the animals to either maximise their breeding or to avoid predation, so there was an advantage to the overall success of the colony.

5.5 Biochemical discussion – what animals may be sensing during lunar events

In a 2015 paper in *Pathophysiology*,[45] Michael Bevington points out that the idea of the moon having effects in plants and animals was proposed in the 1st Century AD by Pliny the Elder,[46] so is certainly not a new idea. However, Bevington suggests that the effects are mainly due to changes in the electromagnetic fields being sensed by organisms. This could be seen as a depression in the melatonin levels.[47] Changes in calcium signalling was also seen, and this is hugely important for a vast range of control of physiological events, including many involved in disease.[48] In both plants and animals, calcium ions are excluded from cells, keeping the internal (cytoplasmic) concentration relatively very low. A significant amount of energy is used by cells to either expel calcium ions across the cell membrane or sequester it (i.e., hide it). By altering these intracellular ion concentrations, the levels of calcium ions can be used to control events in cells, including the activity of enzymes and the release of hormones.[49] Calcium ions control your muscle movements and also the underlying metabolism which allows the chemical energy to be available to drive those muscles. Every time you move, calcium ions are having an instrumental role. Calcium ion concentrations also control the activity of a myriad of enzymes.[50] Therefore, external forces altering this signalling mechanism would be of great significance.

As discussed earlier in Chapter 3, it is thought that many animals may be using the geomagnetic forces to navigate around the globe. Some migrations are truly amazing.[51] The Arctic tern (*Sterna paradisaea*) is reported to migrate from pole to pole, north to south, and back again, a journey of perhaps 40,000 km (25,000 miles) every year. Ackerman suggests that in the lifetime of an Arctic tern, it "may fly the equivalent of three trips to the moon and back".[52] The blackpoll warbler, which is a bird which weighs only approximately 10–20 g, undertakes a migration from eastern Canada and New England to South America. This is a flight of about 1,700–1,800 miles and many of the birds do this non-stop in two to three days. Monarch butterflies (*Danaus plexippus*) migrate to Mexico from Canada and back, a trip of 4,800 km (3,000 miles). In the oceans, Humpback whales (*Megaptera novaeangliae*) migrate approximately 8,000 km (nearly 5,000 miles) each year to search for water dense with food. On land, Wildebeest (*Connochaetes taurinus*) are also famous for migrating, and there are countless other examples. Each of these animals need to be able to find their way, and back again. Famously, salmon (*Salmo salar*) are thought to return to the exact river in which they were born. These fish may use the Earth's magnetic field to find their way, until they can 'smell' the chemicals in the water which matches the river that they are looking for.[53]

Lunar activity, along with the observations of the stars, may be of use for animal navigation too.[54] For example, the yellow underwing moth (*Noctua pronuba*) was reported to use both.[55] In the dark the animals became disorientated. Other insects too seem to use the moon as a cue to orient themselves, such as sandhoppers (*Talitrus saltator*)[56] and dung beetles (*Scarabaeus zambesianus*).[57] However, data such as these reveal that animals may be able to respond to, and even use the moon, but unless much more is known about how this happens it is of little use to us to measure lunar activity or predict stellar events. It should also be remembered that mathematically we can predict the exact motion of the moon[58] and any other planetary activity, so the observation of a few animals may not be of much interest. Even relatively rare events such as the arrival of Halley's comet can be easily predicted. It was last seen in 1986 on its 76-year rotation and will not be seen again until 2061.[59] Therefore, watching the animals for such predictions would only ever be of curiosity, rather than any practical use. On the other hand, a combination of observations may be of major significance. Spring tides can bring disaster.[60] Spring tides arise when the moon and sun align, so that the gravitational pulls on the oceans are at their maximum, and so the tides have the biggest rise. This can bring flooding and is particularly significant if the planets decide to join the party and create what is referred to as super tides.[61] If an animal observation can predict that this will occur simultaneously with bad weather, then a greater disaster may be able to be averted.

So, if animals are susceptible, are humans? Bevington[62] lists many examples where the moon seems to have effects on human health. This includes epilepsy and depression episodes. Gastrointestinal haemorrhage occurrence was increased at the full moon, and even spontaneous abortions seemed to correlate to the lunar cycle. However, using a balanced view, Bevington also lists other effects for which the evidence is weak or lacking altogether. Here, he discusses the perceived increase in emergency care and effects on individuals with bipolar disorder. In a very recent paper, Wehr and Helfrich-Förster[63] start by saying:

Recent longitudinal observations show that human menstrual cycles, sleep-wake cycles and manic-depressive cycles can become synchronized with lunar cycles, but do so in uniquely complex and heterogeneous ways that are unlikely to have been detected by past studies.

This seems quite categoric. Using longitudinal studies, they could show that the several human physiological responses would align with lunar cycles, and concluded that more research is needed in this area. The influence on such synchronisation by light pollution is also posited. Clearly, they felt that there is much more to discover here.

It does, therefore, seem that a range of animals, including humans, can be affected by the moon, and in particular alignment to lunar cycles. However, there is no notion of prediction here, rather the animals are responding to events already happening.

5.6   The solar eclipse, Northern Lights and animals

One of the most amazing natural events which attracts human attention, perhaps as it happens so infrequently, is the solar eclipse. Although some say that it was first referenced in ancient texts about 5,000 years ago,[64] the start of a paper by Sethi says:

> The earliest recorded solar eclipse was on October 22, 2137 BC as mentioned in the Chinese 'Slue Ching'. The Chaldean astronomer in about 400 BC discovered that eclipses occur in regular succession at an interval of about 18 years, the cycle being called the Chaldean Saros.[65]

Regardless of who is right, the unusual nature of an eclipse has led to many cultures being afraid of the event. After all, getting dark in the middle of the day must have been very odd, especially when the sun that you rely on to allow your crops to grow rapidly disappears. It is an odd fact that the moon is exactly the right size, and the exact distance between the Earth and the sun to completely fit one over the other and nothing more. The moon is approximately 400 times smaller, and the sun is approximately 400 times farther away.[66] As ancient peoples saw the moon exactly cover the sun it must have been terrifying, or at least for the few minutes it happened.

On 8th–9th March 2016, in Central Sulawesi, Indonesia, several elements of the environment were monitored and the behaviour of several animal species were observed.[67] Whilst the humidity increased during the eclipse, as would be expected the light intensity, wind speed and temperature dropped. Several animal species were noticeably disturbed, including Heck's macaques, flying foxes, birds (maleo) and amphibians. Several insects showed odd behaviour too. Were any of these predictors of the eclipse? The day before the eclipse, the Black Flying Fox (*Pteropus alecto*) was found to be calling and flying but became still in the roost during the eclipse. They appeared to be responding to the environmental changes, but not predicting it would happen. The Dian's Tarsier (*Tarsius dentatus*) showed no response at all and seemed to simply ignore the event. Domesticated pig (*Sus* sp.) showed abnormal behaviour, whilst the Heck's macaque (*Macaca hecki*) gave out a call which was associated with danger during the darkness. The maleo (*Macrocephalon maleo*) showed no predictive behaviour either but did seem to behave as though it was night during the eclipse. Odd animal behaviour was reported for a range of amphibians, including the Crab eater frog (*Fejervarya cancrivora*), Sulawesian toad (*Ingerophrynus celebensis*), Common Asian toad (*Duttaphrynus melanostictus*) and Chorus frog (*Microhyla* sp.). Several insects were also observed, including the dung beetles (*Paragymnopleurus planus*). However, it did not matter which animal was watched, they all seemed to be undertaking normal behaviour before any environmental cues from the eclipse caused them to be confused. No predictive behaviour was reported.

In an observation of chimpanzees (*Pan troglodytes*), during an eclipse of 30th May 1984, some remarkable animal behaviour was reported. As the eclipse progressed the chimps moved to the top of their climbing frame, and then:

> At 1223 hours, during the period of maximum eclipse, the animals continued to orient their bodies toward the sun and moon and to turn their faces upward. One juvenile stood upright and gestured in the direction of the sun and moon.[68]

The animals were not seen exhibiting such behaviour at other times, such as at sunset. Although not so dramatic, behaviour changes were seen with hamadryas baboons (*Papio hamadryas*) during the 3rd November 1994 eclipse, at the Zoo of Santiago de Chile.[69] However, again, no predictive behaviour was observed in either case.

No predictive behaviour was seen in rats and rabbits in a report during an eclipse in India (16th February 1980, a 70% eclipse at Lucknow).[70] Rats became more active, and rabbits became less, but in response to the event, not before. The effects on the rabbits lasted for 72 hours, but the rats recovered much more quickly, taking about 12 hours. In both species, resorption of embryos was higher if the pregnant animals were exposed to the solar eclipse as well as sunlight. The effect of non-ionising radiation, which is increased during the eclipse, is discussed, but this paper was published in 1980, and although they could only speculate then, more is known now about the effects such radiation may have.[71]

It is not only land animals who may be affected by an eclipse. The gamma rays reaching the surface of the Earth reduce the pH[72] of seawater, and this may in part account for some of the alterations of marine animal behaviour seen. However, again, this is not predictive as it relies on the eclipse happening.[73]

A natural phenomenon which is amazing to see is the Northern Lights, otherwise known as the Arora Borealis.[74] These are caused by the electrical charges emanating from the sun entering the Earth's atmosphere and then being funnelled by the magnetic fields towards the poles. Hence, the phenomenon is only seen relatively near the North or South poles. The energy is absorbed by the molecules (mainly oxygen and nitrogen) in the atmosphere which revert back to what is referred to as their ground state, and as this happens light is given off, which is what we see. Often the aurora is green but red/pink, as well as blue and yellow, are also seen. Interestingly, the human eye is not very sensible to colour under the dark conditions needed to see the aurora well, so we tend to see it as a pale green, almost grey, as the cones[75] of our eyes struggle to be sensitive enough. Photographs and film[76] do not suffer from this colour insensitivity and hence pictures – for example on the television – are quite misleading, showing much brighter colours. Often videos are speeded up too, to show how the aurora swirls, but this movement is in reality very slow. However, to stand on a frozen lake and watch the Northern Lights is an amazing experience, albeit very cold (it was approximately -30 °C in Finland when I did this[77]).

Do animals see or get affected by the auroras? It has been suggested that dogs may look up and be aware of the aurora.[78] It has also been reported that the beaching of whales may be correlated to solar storms and hence the lights.[79] This could be because the whales are using the geomagnetic field for their navigation, and the electrical activity which creates the auroras upsets their senses and makes them effectively 'blind'.[80,81] Losing their navigational skills can lead them into shallow, and dangerous, waters. However, the behaviours of neither the dogs nor the whales are predictive, but rather they are reacting to the aurora.

Animals feature in many myths from around the world which try to explain the Northern lights.[82] For example, some Inuit tribes thought that the lights were the spirits of dead humans which were playing a ball game with a walrus skull. Meanwhile, in Finland, the lights were created by a firefox running so fast that it created sparks, and hence the lights. However, there seems to be no evidence that animals can predict when they will appear. There are apps[83] which can be used for people who wish to see the spectacle. Other Northern Lights predictions include the Aurora Forecast,[84] which uses a ten-point scale to suggest if the lights can be seen. The Geophysical Institute at the University of Alaska Fairbanks also gives a 27-day forecast.[85] A shorter, 30-minute forecast is given by the Space Weather Prediction Center (National Oceanic and Atmospheric Administration).[86] However, even using such forecasting it depends on keeping a regular eye on the sky – the lights seem to come and relatively quickly go – and is of course dependent on having no cloud cover, so a weather forecast is also needed. When I was lucky enough to see the Northern Lights in Äkäslompolo they lasted about 30 minutes. Some places have webcams set to look towards where they are likely to be seen, for example, in Äkäslompolo in Finland.[87] Even so, predicting when it is worth suffering the cold to see the Northern Lights is hard, but animals appear to be no use here.

In summary here, if you wish to see the Northern Lights you need to turn to modern technology and not watch the animals. If you do the latter, you may be too late, as the aurora is already lighting up the sky and you won't have time to pull on all your thermal layers!

5.7 Chapter summary

To conclude this chapter, we will once again return to the solar eclipse and horses. On 21st August 2017, Idaho was in the path of totality and Andrea Maki had her camera phone ready. She was surrounded by 137 wild horses at the Wild Love Preserve. She observed that: "At first, there was a palpable stillness in the air; the horses were on certain alert", whilst during the eclipse the horses started to gallop and then remained spooked for some time. There is even a video online to watch.[88] However, there is little evidence of the horses predicting the event in the footage. The horses certainly responded, but in real time, rather than before. The animals seemed amazingly calm leading up to the light dimming. Perhaps my Cornish horses were spooked by something unrelated to the solar eclipse after all.

On 1st April 1764, a horse was born at Cranbourne Lodge, Windsor, UK, during a total solar eclipse.[89] It was therefore given the name "Eclipse" and then went on to become an immensely successful racehorse. Perhaps its destiny was in some way influenced by the environmental conditions at the time of its birth? It was so successful that many horses subsequently have had "Eclipse" put into their name, obviously hoping that this would help the animal emulate the original. Even if the latter is wishful thinking, the lunar/solar event has been engrained into racehorse culture, cementing a relationship between events above us with animals.

Are animals any use for our prediction of lunar and solar events? It appears that animals do often respond to lunar cycles and other celestial events, but there seems to be no predictive mannerisms which have been noted. It is rather sweet to note that chimpanzees point towards the eclipse, but it must be concluded that such animal behaviour is interesting but of little use as predictive tools. The effects on the birth and health of animals are much more significant and may lead to future discoveries and interventions, so while the whole field of science need not be dismissed, it sadly becomes little more than a footnote for the topic of this book.

Endnotes

1 Avnir, D. (2020) Critical review of chirality indicators of extraterrestrial life. *New Astronomy Reviews*, 101596.
2 Holland, J. (2015) A natural history of disturbance: Time and the solar eclipse. *Configurations*, 23, 215–233.

3 Stephenson, F.R. and Fatoohi, L.J. (1993) Lunar eclipse times recorded in Babylonian history. *Journal for the History of Astronomy*, 24, 255–267.

4 Royal Museums Greenwich: www.rmg.co.uk/stories/topics/history-telescope (Accessed 15/11/21)

5 Space.Com: www.space.com/21950-who-invented-the-telescope.html (Accessed 17/11/21)

6 The BBC: www.bbc.co.uk/history/historic_figures/copernicus.shtml (Accessed 17/11/21)

7 Scientific American: https://blogs.scientificamerican.com/observations/did-galileo-truly-say-and-yet-it-moves-a-modern-detective-story/ (Accessed 17/11/21)

8 Sobel, D. (2009) *Galileo's Daughter: A Drama of Science, Faith and Love.* 4th Estate, London. ISBN: 9781857027129

9 Britannica: www.britannica.com/biography/Johannes-Kepler (Accessed 17/11/21)

10 Britannica: www.britannica.com/biography/Isaac-Newton/Career (Accessed 17/11/21)

11 Holmes, R. (2008) *The Age of Wonder: How the Romantic Generation Discovered the Beauty and Terror of Science.* Harper Press, Toronto, Canada. ISBN: 9780007149537

12 Herschel, W. (1795) XVIII. Description of a forty-feet reflecting telescope. *Philosophical Transactions of the Royal Society of London*, (85), 347–409.

13 The European Space Agency: www.esa.int/Science_Exploration/Space_Science/Herschel/Caroline_and_William_Herschel_Revealing_the_invisible (Accessed 18/11/21)

14 Britannica: www.britannica.com/topic/Mount-Wilson-Observatory (Accessed 22/11/21)

15 Britannica: www.britannica.com/biography/George-Ellery-Hale (Accessed 22/11/21)

16 NASA: https://asd.gsfc.nasa.gov/archive/hubble/overview/hubble_bio.html (Accessed 07/02/22)

17 Interesting Engineering: https://interestingengineering.com/a-brief-history-of-the-telescope-from-1608-to-gamma-rays (Accessed 18/11/21)

18 Jodrellbank: www.jodrellbank.net/visit/whats-here/lovell-telescope/ (Accessed 18/11/21)

19 University of Manchester: www.jb.man.ac.uk/webcam/ (Accessed 18/11/21)

20 Holmes (2008).

21 NASA, Hubblesite: https://hubblesite.org/ (Accessed 18/11/21)

22 NASA, Space Telescope Live: https://spacetelescopelive.org/ (Accessed 30/06/22, but it was finding a new target).

23 NASA, James Webb Space Telescope: https://jwst.nasa.gov/content/webbLaunch/index.html (Accessed 18/11/21)

24 On 22/12/21 the launch was predicted to be delayed by a day because of poor weather (high winds), so would be Christmas Day 2021. The Guardian: www.theguardian.com/science/2021/dec/21/nasas-newest-space-telescope-launch-christmas-day (Accessed 22/12/21)

25 The launch of the James Webb Space Telescope was successful after the day of delay: The Guardian: www.theguardian.com/science/2021/dec/25/nasa-launches-james-webb-space-telescope (Accessed 05/01/22)

26 Webb Telescope: https://webbtelescope.org/news/first-images/gallery (Accessed 05/08/22)

27 The European Space Agency: www.esa.int/Science_Exploration/Space_Science/Herschel_overview (Accessed 19/11/21)

28 W.M. Keck Observatory: www.keckobservatory.org/ (Accessed 19/11/21)

29 University of Hawai'i at Hilo: www.malamamaunakea.org/articles/9/Maunakea (Accessed 19/11/21)

30 Five-hundred-meter Aperture Spherical radio Telescope: https://fast.bao.ac.cn/ (Accessed 19/11/21)

31 A depression in the ground caused by the dissolution of soluble rocks.

32 Gran Telescopio Canarias: www.gtc.iac.es/GTChome.php (Accessed 19/11/21)

33 Science Focus: www.sciencefocus.com/space/what-is-the-biggest-telescope-on-earth/ (Accessed 19/11/21)

34 Sullivan, J.D., Ditchkoff, S.S., Collier, B.A., Ruth, C.R. and Raglin, J.B. (2016) Movement with the moon: White-tailed deer activity and solunar events. *Journal of the Southeastern Association of Fish and Wildlife Agencies*, 3, 225–232.

35 Survival Freedom: https://survivalfreedom.com/what-does-moon-overhead-and-underfoot-mean/?utm_source=rss&utm_medium=rss&utm_campaign=what-does-moon-overhead-and-underfoot-mean (Accessed 19/11/21)

36 Murray, J. (1996) *Moon Struck; Hunting Strategies That Revolve around the Moon.* Adventure Pubns, Cambridge, MN. ISBN: 9780964682306

37 Murray, J. (2004) *Moon-Phase Deer Hunting.* Creative Outdoors. ISBN: 9781580112178

38 Navas González, F.J., Jordana Vidal, J., Pizarro Inostroza, G., Arando Arbulu, A. and Delgado Bermejo, J.V. (2018) Can donkey behavior and cognition be used to trace back, explain, or forecast moon cycle and weather events? *Animals*, 8(11), 215.

39 SciJinks: https://scijinks.gov/tides/ (Accessed 19/11/21)

40 Navas González *et al.* (2018).

41 Yonezawa, T., Uchida, M., Tomioka, M. and Matsuki, N. (2016) Lunar cycle influences spontaneous delivery in cows. *PLoS ONE,* 11, e0161735.

42 Zimecki, M. (2006) The lunar cycle: Effects on human and animal behavior and physiology. *Poste̜py Higieny i Medycyny Do'swiadczalnej*, 60, 1. [The Abstract is in English]

43 Takemura, A., Rahman, M.S. and Park, Y.J. (2010) External and internal controls of lunar-related reproductive rhythms in fishes. *Journal of Fish Biology*, 76(1), 7–26.

44 Grant, R.A., Chadwick, E.A. and Halliday, T.R. (2009) The lunar cycle; a cue for amphibian breeding phenology? *Animal Behaviour*, 78, 349–357.

45 Bevington, M. (2015) Lunar biological effects and the magnetosphere. *Pathophysiology*, 22(4), 211–222.

46 Pliny the Elder, Naturalis Historiae [On Natural History]: Lucilius: 2.41; penetrating all things: 2.102; 8.80; Mucianus: 8.20; Pytheas 2.19. Bostock, J. and Riley, H.T. (eds.) (1855) *Pliny, the Elder: The Natural History.* Taylor and Francis, London [reference given as cited by Bevington (2015)]

47 Palmer, S.J., Rycroft, M.J. and Cermack, M. (2006) Solar and geomagnetic activity, extremely low frequency magnetic and electric fields and human health at the Earth's surface. *Surveys in Geophysics*, 27(5), 557–595.

48 Patergnani, S., Danese, A., Bouhamida, E., Aguiari, G., Previati, M., Pinton, P. and Giorgi, C. (2020) Various aspects of calcium signaling in the regulation of apoptosis, autophagy, cell proliferation, and cancer. *International Journal of Molecular Sciences*, 21(21), 8323.

49 Dodd, A.N., Kudla, J. and Sanders, D. (2010) The language of calcium signaling. *Annual Review of Plant Biology*, 61, 593–620.

50 Hancock, J.T. (2021) *Cell Signalling*. Oxford University Press, Oxford. ISBN: 9780198859581 [section 4.7 in particular].

51 National Geographic: www.nationalgeographic.org/article/natures-most-impressive-animal-migrations/ (Accessed 22/11/21)

52 Ackerman, J. (2016) *The Genius of Birds*. Corsair, London. ISBN: 9781472114365

53 US Geological Survey: www.usgs.gov/faqs/how-do-salmon-know-where-their-home-when-they-return-ocean-1?qt-news_science_products=0#qt-news_science_products (Accessed 22/11/21)

54 Foster, J.J., Smolka, J., Nilsson, D.E. and Dacke, M. (2018) How animals follow the stars. *Proceedings of the Royal Society B: Biological Sciences*, 285(1871), 20172322.

55 Sotthibandhu, S. and Baker, R.R. (1979) Celestial orientation by the large yellow underwing moth. *Noctua pronuba L. Animal Behaviour*, 27, 786–800.

56 Papi, F. and Pardi, L. (1963) On the lunar orientation of sandhoppers (*Amphipoda Talitridae*). *The Biological Bulletin*, 124(1), 97–105.

57 Dacke, M., Byrne, M.J., Scholtz, C.H. and Warrant, E.J. (2004) Lunar orientation in a beetle. *Proceedings of the Royal Society of London. Series B: Biological Sciences*, 271(1537), 361–365.

58 For example, lunar activity can be determined by such tools as this at Star Date: https://stardate.org/nightsky/moon (Accessed 05/21/22)

59 EarthSky: https://earthsky.org/space/halleys-comet-and-edmond-halleys-prediction/ (Accessed 05/01/22)

60 This from my home city of Bristol. BristolLive: www.bristolpost.co.uk/news/bristol-news/gallery/pictures-high-spring-tides-flood-3942718 (Accessed 05/01/22)

61 The Environment Agency: https://environmentagency.blog.gov.uk/2015/02/20/super-tides-the-weather-and-coastal-flood-risk/ (Accessed 05/01/22)

62 Bevington (2015).

63 Wehr, T.A. and Helfrich-Förster, C. (2021) Longitudinal observations call into question the scientific consensus that humans are unaffected by lunar cycles. *BioEssays*, 2100054.

64 Live Science: www.livescience.com/60139-why-eclipses-frightened-ancient-civilizations.html (Accessed 22/11/21)

65 Sethi, N. (1980) Behavioural and tetratogenic effects of solar eclipse. *Indian Journal of Psychiatry*, 22(4), 390.

66 Scientific American: https://blogs.scientificamerican.com/life-unbounded/the-solar-eclipse-coincidence/ (Accessed 22/11/21)

67 Wiantoro, S. (2019) Effects of the total solar eclipse of March 9, 2016 on the animal behaviour. *Journal of Tropical Biology & Conservation (JTBC)*, 16, 141–153.

68 Branch, J.E. and Gust, D.A. (1986) Effect of solar eclipse on the behavior of a captive group of chimpanzees (*Pan troglodytes*). *American Journal of Primatology*, 11(4), 367–373.

69 Gil-Burmann, C. and Beltrami, M. (2003) Effect of solar eclipse on the behavior of a captive group of hamadryas baboons (*Papio hamadryas*). *Zoo Biology: Published in Affiliation with the American Zoo and Aquarium Association*, 22(3), 299–303.

70 Sethi (1980).

71 Mallery, E. (2020) Consensus statement of UK and International Medical and Scientific Experts and Practitioners on Health Effects of Non-Ionising Radiation (NIR). https://stopsmartmetersbc.com/wp-content/uploads/2020/11/2020ConsensusStatementOfUKAndInternationalMedicalAndScientificExpertsAndPractitionersOnHealthEffectsOfNon-IonisingRadiationAndRadiofrequency

Radiation-Dr.-Erica-Mallery-Blythe.pdf-IonisingRadiationAndRadiofrequen cyRadiation-Dr.-Erica-Mallery-Blythe.pdf (stopsmartmetersbc.com) (Accessed 22/11/21)

72 pH is a measure of acidity, i.e. H⁺ content. The lower the pH the more acid the solution, the scale going up to 14. The inside of cells, and most biological solutions, are around neutral, pH7.

73 Kumar, S.S. and Rengaiyan, R. (2014) Vedic mythology of solar eclipse and its scientific validation. *Indian Journal of Traditional Knowledge,* 13(4), 716–724.

74 The *Aurora Borealis* is seen in the northern hemisphere, whilst in the southern hemisphere the equivalent is the *Aurora Australis* (Southern Lights).

75 The cells of the eye sensitive to light are the rods and cones. Rods are used for general light, but the cones are used for colour perception. However, they are less sensitive than rods.

76 Even camera phones can show spectacular colours if they can be altered to high ISO and long exposure settings.

77 My camera-release wires and the plastic centre of my camera tripod both snapped off, I assume because of the temperature. Later I was warned that carbon-fibre tripods become brittle at these temperatures and should also be avoided – I do not know if this is true.

78 Hurtigruten: www.hurtigruten.co.uk/inspiration/experiences/northern-lights/faq/ (Accessed 07/12/21)

79 Labroots: www.labroots.com/trending/plants-and-animals/6835/link-northern-lights-whale-beachings (Accessed 07/12/21)

80 Granger, J., Walkowicz, L., Fitak, R. and Johnsen, S. (2020) Gray whales strand more often on days with increased levels of atmospheric radio-frequency noise. *Current Biology,* 30(4), R155–R156.

81 Forbes: www.forbes.com/sites/jamiecartereurope/2020/04/03/how-the-northern-lights-make-whales-go-blind-and-get-lost/?sh=3b6c7c8278b8 (Accessed 07/11/21)

82 Aurora Zone: www.theaurorazone.com/about-the-aurora/aurora-legends (Accessed 07/12/21)

83 For example: *Aurora Alerts – Northern Lights forecast*

84 Aurora Forecast: https://auroraforecast.com/#:~:text=3%2DDay%20Forecast&text=Aurora-,Active%20aurora%20possible%20with%20good%20chance%20of%20isolated%20minor%20auroral,isolated%20major%20aurora%20substorms%20possible.&text=Aurora-,Active%20aurora%20possible%20with%20good%20chance%20of%20isolated%20minor%20auroral,conditions%20for%20major%20aurora%20activity (Accessed 03/08/22)

85 Geophysical Institute at the University of Alaska Fairbanks: www.gi.alaska.edu/monitors/aurora-forecast (Accessed 03/08/22)

86 Space Weather Prediction Center (National Oceanic and Atmospheric Administration): www.swpc.noaa.gov/products/aurora-30-minute-forecast (Accessed 03/08/22)

87 Ylläs Log Cabins: http://yllaslogcabins.com/ (Accessed 07/12/21)

88 Horsetalk: www.horsetalk.co.nz/2017/09/07/total-eclipse-unsettling-wild-horses/ (Accessed 19/11/21)

89 America's Best Racing: www.americasbestracing.net/the-sport/2021-eclipses-important-part-racing-history (Accessed 19/11/21)

6

Conclusions, climate change and the future

6.1 Introductory story

George had been watching his Tempest Prognosticator for days. The little leeches were happy in the rainwater at the bottom of their jars. Despite the fact that he had made sure that they could see each other, having the jars made of glass and arranged neatly in a circle, the little animals never seemed to notice one another, or anything else for that matter. George went to bed disappointed.

The next morning the leeches were halfway up the jars. However, the ones on the east side of the machine were much higher than the others. This was the side that was facing the sea. A storm was coming, and directly in from the North Sea. It could be bad. As he watched, two of the leeches reached the top and the bell rang. It was time to move. The machine had given a clear indication. Despite all the derision he had faced, he was sure that now was the time to warn the town.

George ran to the Town Hall, his jacket flapping as he rushed along. It was early, so he banged on the door. Nothing stirred, so George thumped on the door harder. Still nothing. Still determined, but not willing to do any damage, George rapped on the door with the brass end of his stick. Surely, that could be heard from the offices inside.

"What?" a voice shouted, and then the door cracked open.

"There's a storm coming. Straight from the sea, and it's going to be bad."

"Not you again? Go away. It's early. Too early for this nonsense."

"But the machine-"

"Your leeches talking again? Go away, George."

"Please, I implore you," George tried.

"Go back and check your barometer," the man said.

George trudged back up the street, despondently. He had had great plans for the weather forecasting machine, but no one would take him seriously. He hoped to make

DOI: 10.1201/9781003343844-6

a large number of them and have them installed in all major towns and cities. If only he could successfully predict bad weather and save a lot of lives? Then he would be taken seriously. They would have to listen to him if that happened. He would persevere and one day people would take him seriously. But today was not going to be the day.

Back in the house, four leeches had made the climb to the top of the bottles and had rung the bell. George carefully dropped them back into the water at the bottom of the flasks and reset the whale bones. It would have to wait until another day for it to be celebrated. However, George was sure that that day would come.

The date was 1850. That evening a "Great Storm" hit the east coast of the UK. Of course, George and the Prognosticator were given no credit for predicting its arrival and he was left to celebrate alone with his leeches.

6.2   The environment – an introduction

There is no doubt that our environment is vastly important to us. We rely on the ground to support our crops and for the weather to water them. We also rely on the ground not erupting and causing huge amounts of damage.

Throughout this text has been the underpinning notion that we can predict changes that may be on the way using the behaviour of animals. However, such predictions are not always needed. Some regions of the world have the same weather, day in and day out. The observations of animals at such places would be fruitless.

I once had the amazing privilege to be hovering in a helicopter. We had taken off from the airport at Maui, Hawaii, and flown on a pleasure trip to see the island. The island is made up of two volcanoes and has the nickname of "Valley Island". It was whilst flying over the more western side of the island when the pilot stopped and hovered. He then said that if we looked to the left, we would see rainforest. If we looked to the right, we would see desert. He told us, and I only have his word for this, that this was the only point on earth where one could observe rainforest and desert at the same time. The weather on those sides of the mountain never changed.

On an earlier trip to Hawaii, I had sat on Waikiki beach and watched the clouds. They ominously hung in the sky to the north. You could see them moving, but they never arrived. Rather stupidly, we then hired an open-topped jeep and toured around the island. We headed past Diamond Head and up the eastern coast. And then it started to rain. We had no roof, and we were soaked very quickly. On the other hand, we were young and on holiday, so it did not really matter, as long as we did not damage the jeep, and it was built for it. So, on we went, through the rain, observing the wonderful surfing beaches on the north coast before coming around the west of the island, and back into bright sunshine. The clouds still moved towards Waikiki, but they never arrived. The weather pattern of the island is like that nearly every day.

This is typical of oceanic islands. Winds pick up the moisture from the sea surface and blow across the mountainous islands. The air rises and cools, clouds form and rain results. As the air drops over the other side of the volcano the clouds magically disperse and no rain falls. It could be like that for weeks. Hence, on Maui, one side of the mountain sustains the rainforest whilst on the other side is a desert. It has been like that for so long that it is completely predictable. No animals needed here, or indeed any other weather forecasting. The locals know what to expect, every day.

Back in Cornwall we sometimes had a similar weather pattern. When younger I would love to go to the beach, and we had an embarrassment of choice. The softer and safer coast on the south, or the more rugged and a little more dangerous north coast, where the surfing was much better. I can distinctly remember leaving the sunny south coast, reaching the ridge of the county only to find fog and rain on the north coast. We would turn around and head home. However, such weather is hard to predict in the UK. Here, good forecasting is needed, and perhaps the animals could help?

It is not only relativity benign weather phenomena that are predictable. In the USA, there is a region which has the nickname of "Tornado Alley". Here, there is a season where tornadoes are likely to happen (colloquially known as twisters), and there is what is referred to as a tornado season. In the Midwestern states this is usually in the spring, but it varies depending where in the country you are. However, there are reports that the periods of peak of activity are shifting, although the authors say that it is hard to attribute this to long-scale climate change.[1] Tornadoes will happen on a frequent basis and as such people go tornado chasing. Companies even run tours so you can go on holiday to chase them yourself.[2]

No matter how predictable we think the weather, the world is changing. The global average temperature is rising. Ice is melting, and sea levels are rising. Some think that it is too late to stop climate change. In his recent book,[3] Bill McGuire has a very negative view. In his review of McGuire's book, Dr Stuart Parkinson[4] says that McGuire "points out just how little time we have left to stop the climate crisis engulfing human civilisation". Although such phrasing sounds apocalyptic, small changes are still going to impact on the weather and other environmental factors.

As I sit and write this book, I recall that on several occasions in the last few days I have said to friends and colleagues that it is a late autumn. The deciduous tree outside my window is still green, with hardly a hint of the usual autumnal oranges and reds, and it is the start of November (in 2021).

A Twister in Wyoming, United States.
(Photo by Nikolas Noonan on Unsplash. Image made black and white.)

This is very unusual for the southwest of England. The leaf clearing rota at a local sports club has been cancelled twice already. The rakes are propped ready, but the trees are in no rush to drop their leaves. Interestingly, spring 2022 was said to have come very early in the UK,[5] and it is thought that will now become the norm. Signs of an early season include plants flowering and the sighting of a beetle, being the earliest such an event that has ever been recorded. Now, as I continue to edit this text, the UK is breaking heat records, with temperatures being recorded above 40 °C for the first time ever.[6] July was the driest for England since 1935 and for some regions of the UK, such as the southeast, July 2022 was the driest on record[7] – records began in 1836. As I look today (early August 2022) my internet weather forecast shows little indication of any rain in the next couple of weeks. But this is the UK, and we always moan about rain over the summer – not in 2022 it seems. It appears as though climate change, as would be expected, is going to have long-term and lasting effects.

6.3 Differences of weather impacts around the world

Although the weather is an obsession in the UK,[8] observing our environment is important for most cultures around the world. We do not just observe either but use this as a judgement of the weather or the climate more generally. The weather often dictates our activities. This is as important in the UK as anywhere else. Do I put the washing on the line, or is rain on the way? Shall I stay at home as snow is forecast? Shall I go to the beach for a swim, or is it going to be too rough and cold? The weather, and the forecast of future weather, can have a profound influence of human activity. Long-term weather prediction can determine the timing of crops being sown, crops being harvested or the escape from a devastating storm. Even though humans have developed ever more sophisticated and expensive technology to predict the weather (as listed in Table 3), and the environment more widely, many animals seem to be able to do the same in an intuitive manner (as listed in Table 1).

It needs to be considered, however, what the impact of weather is to different regions and the people who live there. In the UK we like to moan about yet more rain. Others around the world will be hoping rain will come. Cloud seeding in some countries tries to encourage rain, hardly a thing we need to do in the UK (although August 2022 may make us think otherwise with hose-pipe bans being announced). However, if crops have not seen rain too long, then precipitation is needed and this would be seen as a good thing. This is important when trying to interpret how the behaviour of animals is regarded in many cultures, and in the literature which is cited, here and elsewhere. For many, the arrival of a good season may be the coming of rain. At looking at the examples in Table 1, it is not surprising perhaps, that so many examples of animal observations are to do with rain and storms. To some, rain brings health to their environment, where crops can thrive and rivers flow deeply, encouraging fish and other animals to prosper. To others, the coming of heavy rain and storms could bring destruction. Crops may be damaged and may not be able to be harvested. Wet crops may rot, rather than store well. Severe storms may cause infrastructure damage, and delay or stop transport routes.

Care needs to be taken, therefore, when interpreting how the observation of animals may be helping. What is good for one, is not necessarily good for others. No one wishes to be washed away by a tsunami, or be crushed by an earthquake. Some predictions are useful for everyone, and if animals can help here then all the better.

6.4 Are animals of use as environmental predictors?

The start of this book told a story of horses, potentially spooked by an oncoming eclipse. Were those horses sensing something I was unaware of at the time? Perhaps. In the earlier chapters there are numerous examples of where animals are used by people to determine the onset of weather patterns. This more commonly seems to be arrival of rain, but not always. Even long-range forecasting can be determined by watching the behaviour of the right animals. For example, C. Makwara Enock suggested that the observation of ants may aid farmers to prepare for an impending storm, whilst the observation of storks may indicate a good season ahead.[9] This is important for many societies around the world, certainly not just in the UK. When do I plant or harvest crops? Should I batten down before a storm, or even move away altogether? Survival of such societies depends on such decisions, and those decisions need to be informed by something. And before you are too sceptical, remember that such societies were in existence, and have survived, long before the invention of the thermometer, or the barometer or any hint of modern weather forecasting using satellites and computers. Many societies live where impending doom may just be around the corner, be that a strip of low-lying coastal land or on the side of a volcano. And many cultures still do not have access to high-tech toys to determine if the rains are coming, or how long they may persist. Watching and understanding our environment is important. Of course, animals and their observation are just part of wider knowledge system, but it is an important aspect in many cultures. "Uncle Offa" mentions many examples of animals and weather forecasts, and even earthquake prediction, but he also uses many other aspects of peoples' cultures as environmental predictors, mainly which day of the year events happen. For example, he points out that in Norfolk (a county in the east of England) it is said that:

If Christmas Day on a Thursday be
A windy winter we shall see.

(Norfolk)[10]

His book gives numerous other examples, some which are hard to comprehend, but clearly, they seem embedded in cultures long enough to endure. A full discussion of these is outside the scope of the discussion here, but these sayings are interesting to note.

Animals are not just useful for short-term weather forecasting either. We are in a time of unprecedented climate change, and this could have devastating consequences for the flora and fauna around us. Watching how our environment adapts to changes in conditions may be crucial to how we make future decisions, and this will involve watching animals as well as plants.

It has been suggested that as the climate changes, plants will not be able to move fast enough. We are not talking triffids[11] here, but the movement of a species. As the climate changes, it is likely that regions of the world will become unsuitable for certain species to thrive. However, as the climate warms, neighbouring areas may become suitable and species might be able to survive in places where not previously possible. If the climate changes so fast that those new areas become unhabitable for that species before they can get established, the plant may completely die out in that region. Those plants may be crucial to the survival of the whole regional ecosystem, so the loss of just that plant may have much greater repercussions than first imagined and there may be significant effects on other plants and animals. Corlett and Westcott[12] discuss the velocity of movement of plants and the speed of climate change to assess if this is really an issue. They think this is of particular importance and say: " 'Migration lag' is a particular concern with plants, because it could threaten both biodiversity and carbon storage." It does seem from such discussions that if the climate changes too quickly, some plants will not continue to thrive, and may disappear from regions of the globe. Viana[13] argues that the long-range dispersal capacity of aquatic plants also needs to be understood, so that the potential of the loss of such species can be estimated as climates change. It has also been suggested that the alteration of migration patterns of animals, caused by changing climates, may also affect the survival of some plant species, with possible extinction being the result.[14] Plants, therefore, may be particularly susceptible to climate change, much more than we might have thought.

Would it be the same for animals? This would not be a concern with many animals who can simply walk or crawl to new habitats, but it may be an issue with animals who are more sessile. An elephant can walk a long way, but some microscopic animals may not have the luxury of wandering off to new climes when times become hard. Some animals are instrumental in allowing agriculture to continue. Nematodes are a good indicator of soil quality, for example.[15] In related work, it was found that the changes in environmental conditions did alter the soil microbiological ecosystems, and the authors stated: "However, biodiversity was generally affected negatively by the environmental manipulations."[16]

Even if animals can move, they may not be immune to being trapped by climate change. On Mount Karimui, a peak of nearly 3,000m in Papua New Guinea, it has been reported that the bird-of-paradise has ascended about 300 feet as it looks for cooler air as the climate warms. As pointed out,[17] mountains are like pyramids, and there is less land nearer the top, and eventually the land runs out altogether. The white-tipped robin has now

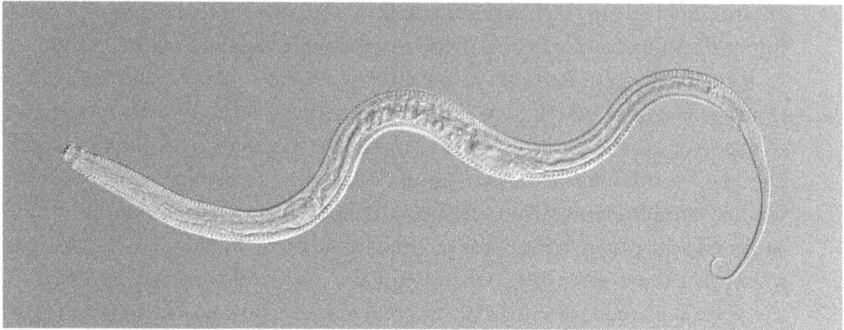

A nematode seen under phase contrast microscopy. Although there is no scale given here, nematodes such as *Caenorhabditis elegans* can be seen with the naked eye but only just.
(Sourced from Shutterstock, contributed by Hussmann, but made black and white.)

been squeezed into the top 400 feet of the mountain. Other bird species are affected too, and if the temperature in New Guinea rises as projected (+4.5 °F), these birds will run out of ecosystem to inhabit.

Of relevance here is that as the climate changes, the observation of the environment is of vital importance. As plants and animals find it harder and harder to live in certain regions of the world there is a potential for the loss of indigenous knowledge systems. Some knowledge has been embedded in many cultures for generations, and has endured, giving guidance to how the local people need to behave for their survival. Changes to, or failure to understanding future changes of, such knowledge would be a great loss to many societies.

It is, therefore, important that indigenous knowledge systems, including the use of animals, is not lost from cultures. For example, Eira *et al.*[18] argue that the knowledge of how reindeer are herded through snow is important to understand how the conditions in the arctic are changing, and this knowledge comes from the local herdsmen. As animals are forced to move, alter their migration patterns, valuable knowledge of our environment will be lost. Worst still, animals may become extinct. Cahill *et al.*[19] say:

> Collectively, these results highlight our disturbingly limited knowledge of this crucial issue but also support the idea that changing species interactions are an important cause of documented population declines and extinctions related to climate change.

Once an animal becomes extinct, it will never come back. Even genomic[20] sequencing, DNA amplification and cloning is unlikely to restore an animal once lost. For years there has been talk about bringing back the mammoth, but it is certainly not easy.[21] In some cases, restoring extinct animals may not actually be desirable. It made a good story in *Jurassic Park*, but that film also made it clear that 'meddling with nature' was not a good thing to do for a variety of reasons[22] and Richard Attenborough's character was cast as being a bit of a madman.[23] It seems more likely that if animals, or plants, become extinct, vital environmental knowledge gained by observing that animal (or plant) will have been lost forever too.

Of course, animals are not the only markers of the arrival of different weather. Many cultures use a range of indicators, as was mentioned in Chapter 3, with many researchers taking an indigenous knowledge systems approach.[24,25,26] This would involve observations of the stars and the moon,[27] as well as another environmental cues, such as the local plants. Perhaps physicists such as Shapley were not too far off the mark by studying both the stars and ants. With a different slant on the use of the stars, astrology has been used for weather forecasting, amongst many other things such as medicine, since the Medieval periods.[28] In popular works, a good example of watching your environment is seen in Delia Owens's debut novel[29] *When the Crawdads Sing*.[30] Owens is an American zoologist and conservationist, working particularly in Africa.[31] In her book, which has just been released as a film,[32] one of the characters, Kya's father, explained "how to read the weather in the clouds and the riptides in the waves". This has significance for the end of the story, which of course I will not spoil for you.

6.5 Observation of plants as environmental predictors

Even though the focus of this book is on animals, it would be amiss to not consider plants in this look at organisms which can be used for forecasting or measuring our environment. Plants are particularly useful in the toolbox of weather prediction.

The observations of plants, like that of animals, is not a new idea. In a 2003 paper by Lantz and Turner[33] they say:

> When Samuel de Champlain arrived at Cape Cod in 1605, the Wampanoag people informed him that the best time to plant corn was when the white oak (*Quercus alba* L.) leaf was the same size as the footprint of a red squirrel (*Tamiasciurus hudsonicus* Erxleben).

Although this mentions an animal, the squirrel is somewhat irrelevant, as it is just being used as a scale. However, the quote shows that the observation of plants was being used to alter human activity, and activity that was important. It would not be good to plant the corn if inappropriate conditions were on the horizon.

Lantz and Turner also talk about other examples, but here the plants were being used as an indicator of whether the animals were good to eat. This includes:

> the Haida utilization of the blooming of cow parsnip (*Heracleum lanatum* Michx.) as a sign that seagull (*Larus* spp.) eggs were no longer good to harvest (Turner 1998a); and the Okanagan use of mock-orange (*Philadelphus lewisii* Pursh) blooming as an indicator that the marmots (*Marmota* spp.) were fat and ready to be hunted.

Several other examples are also discussed. In western Canada fishermen watch for when the southern cottonwood (*Populus balsamifera* L.) releases seeds to indicate that it is a good time for pickerel (*Esox lucius* L.), whilst on the other side of the country fishermen are watching for the flowering of the saskatoon (or shadbush,[34] *Amelanchier* spp.) before they fish for shad (*Alosa sapisissima* Wilson). Clearly, watching the plants for determining future human activity is important.

With a more direct focus on the prediction of future weather the *Almanac*[35] lists numerous examples of how the observation of plants can be used to forecast. This list starts with the pinecone, which will be open and be stiff if good weather is coming, but rising humidity, indicating the onset of rain, will be indicated by the cone becoming softer and closing. I remember my parents back in Cornwall telling me the same thing whenever I picked up a cone in the woods. Other indicators of rain are given in the *Almanac*, including oak and maple leaves curling as the humidity rises, the closing of the flowers of common chickweed (or stitchwort: *Stellaria media*), as well as the purple sandwort and the scarlet pimpernel. Apparently, the pimpernel is known as the "poor man's weatherglass",[36] and led to this little saying:

> Pimpernel, pimpernel, tell me true. Whether the weather be fine or no. No heart can think, no tongue can tell. The virtues of the pimpernel.[37]

The *Almanac* goes on to say that the timing of the opening of the African marigold is important, as if it opens later in the day or closes early, rain may be on the way. On the other hand, opening of the flowers of morning glory indicate that the weather is going to be fair.

"Uncle Offa"[38] also has an opinion on plants: "an abundance of acorns, dead nettles and thick onion skins in October threaten a hard winter", for example. He also suggests that sycamore or poplar will "reveal the lighter-coloured underside of their leaves" if rain is on the way.

Such listings about plants as seen in the *Almanac* or by "Uncle Offa" show, like that for animals, there are numerous examples of how the observation of plants can be used to forecast the weather, often on a short time scale, as the changes seen are a response to changing conditions which are already happening, such as humidity. It also shows how such observations, again like that for animals, are well embedded into cultures and how they have lasted the test of time.

There are numerous other examples available elsewhere too. On a website highlighting Māori culture,[39] used earlier when animals were discussed, there are several interesting examples of plants being useful. These include the early flowering of the Tī kouka (Cabbage tree) which indicates that there will be a long hot summer. Periodic blooming of the Pōānanga (Clematis) shows that the season will be warm and have gentle breezes. With the Puahou (Five finger), it depends on which part of the plant changes: if the plant blooms at the top and the flowers progress down the plant a cold and bad season is on the way, whereas if it is the other way around a warm season is coming.[40]

Another good source of information about the observing of plants for weather forecasting is a paper by Anely Nedelcheva and Yunus Dogan,[41] who looked at the use of plants in Bulgarian folk traditions. Although published in 2011, the study weas carried out in 2006–2007, and they looked at 30 species of plants, with 20 of those being commonly featured. They list the observations, including flower blooming, fruit and other crops, and leaf fall. The list is too long to use fully here, but there are some notable examples worth pulling out, including, if spring beans grow a lot, or apricots are abundant, the season will not be a good one, i.e., "year will not be fertile", and "When wild apples bear much fruit, the winter will be *evil*." [their italics here].

The conclusion of this paper is worth pausing over. The authors state:

> Traditional plant-based weather and climate forecasts are still passed along from generation to generation. Although the plants have lost some of their importance in weather forecasting and predicting natural disasters in the modern world, their study helps to reveal the mechanisms of establishing folk botanical knowledge as part of a country's national characteristics, and to differentiate its rational basis, and it is source of empirical data on phenology and plant physiology.

And they conclude with:

> This kind of study is opportune at the present time when, more than ever before, the world is concerned with climatic change and its impact on humans and our environments.

> [remembering this was written in 2011, and
> it is even more pertinent now in 2022]

The observation of the activities of plants, i.e., whether they flower, give good crops et cetera, are certainly worth putting in the mix along with looking at the mannerisms of animals, alongside with all the other factors which make up a holistic predictive system. As the climate changes the observations of plants, if they thrive or die, will certainly be important to enable us to understand how our environments are being affected. After all, crops from plants are the sustenance all animals need (even if they are high up in the food chain, plants were needed lower down), and as humans are persuaded to reduce their meat consumption,[42] because this will reduce greenhouse gas emissions, plants will simply become even more important.

We started the main chapters of this book by discussing if animals can be used as a thermometer, with Shapely showing that the speed of an ant could be used to measure ambient temperatures. According to the *Almanac*,[43] plants might have a use here too. The leaves of the Rhododendron will open more as the temperature increases. As stated in the *Almanac*: "They will be closed at 20 degrees by when the temperature is above 60, the leaves are open!" Obviously, the temperatures here are being quoted in Fahrenheit. Others too have written about this and suggest that it might help the plant survive freezing and thawing cycles in the winter.[44] The phenomenon has even got into the scientific journals.[45] Perhaps Shapley should have looked at the plants as well as his beloved ants. Having said that, Rhododendrons are extremely common in Cornwall, having been imported by the plant hunters of past centuries[46] and then accidentally spread along the railway lines and planted in a myriad of gardens, including my own childhood one. I do not remember ever been told to watch the Rhododendrons before deciding to put on a jacket.

6.6   Use of sentinel organisms for environmental monitoring

The observations of many organisms have been used for environmental monitoring for years. Many species are described as sentinel organisms. One definition given of such organisms is: "Sentinel species, which readily accumulate pollutants and therefore serve as indicators of ecosystem health."[47] These are therefore species which can give an indication of the long-term changes in an environment. Such environmental changes may be due to anthropogenic activity. Factories may be discharging waste into water courses, and for this the health of the water may be monitored by observing how sentinel species are coping. Are they unhealthy or simply disappearing?

There are numerous examples in the scientific literature on the use of sentinel organisms. For example, blue mussels (*Mytilus edulis* spp.) are being used to monitor coastal pollution,[48] along with fish[49] and snails.[50,51] Even the parasites that live on other organisms can be used,[52] or organisms can be used to estimate the health of soils.[53]

This is not predicting the imminent nature alterations to the environment brought on by the changing of the weather or an imminent earthquake. This type of observation is often long-term and in response to a specific need. However, as the climate changes the use of environmental observations will becomes more important. Some suggest looking at whole ecosystems, such as lakes,[54,55] whilst others suggest monitoring certain species or organism groups, such as marine mammals,[56] or birds.[57]

The work with sentinel organisms will involve their observation, but also how they adapt to change and how their biochemical processes are impacted on. It is likely that sentinel organisms will show stress responses and subsequently there will be changes in the genes which they express. For example, see the work of Ferreira *et al.*[58]

However, a full discussion of the use of sentinel organisms is not part of the argument being put forward here. It is undoubted that sentinels are hugely important and that monitoring them will be of more significance in the future, but they are not being used to predict the weather, or whether a tsunami is about to roll in. Therefore, this brief overview of their role in environmental monitoring will end here.

6.7 Some of the famous who have dabbled

The roam through the literature about using animals for environmental forecasting has been dependant on the writing of many others over the course of history. In this light, it is interesting to reflect on the work of some of the scientists and other eminent people encountered in the preceding chapters. Many were investigating, theorising and dabbling in a wide range of topics, not only those for which they became famous. Harlow Shapley, a famous astronomer, also studied the speed of ants. Amos Dolbear, a famous physicist who invented early versions of the radio and telephone, published on the sound of crickets. Edward Jenner, the inventor of vaccines, and so the person who has saved countless lives (including during the present COVID-19 pandemic), also thought about animals and the weather. He was said to have looked at spiders, and he said that if they crawl out whilst it is still raining the weather will soon improve, the rain being short-lived and light.[59] According to Ewer,[60] Jenner was also said to have given this quote:

> If the cat washes her face o'er the ear 'Tis a sign the weather'll be fine and clear.

Ewer then goes on to say: "It seems that this most independent of domestic animals has refused to serve man even as a reliable barometer."

Of significance here, Jenner also wrote about the antics of cuckoos and inspired Merryweather's Tempest Prognosticator, and this was because Jenner also wrote poetry.

Other eminent thinkers and writers also gave interesting quotes, again from Ewer's paper:

> Thus Jonathan Swift tells us that 'Careful observers may foretell the hour (By sure prognostics) when to dread a shower; While rain depends, the pensive cat gives o'er Her frolics and pursues her tail no more.'

Mind you, Swift is also supposed to have said: "Tis very warm weather when one's in bed."[61] It has to be assumed that the Jonathan Swift that Ewer is quoting is the same one and is the Irish writer and political commentator (1667–1745)[62] who wrote *Gulliver's Travels* (1726), amongst lots of other things. Ewer also quotes John Clare and says: "considered that a storm would follow if 'the cat with her tail runneth round till she reels'". I assume that this is John Clare[63] (1793–1864) the poet, although in the copy of Ewer's paper which I could access there were no citations given. However, "Uncle Offa" refers too to John Clare, describing him as a "Northamptonshire poet".[64]

Even the scientists' assistants, such as Margarette W. Brooks, got in on the act. These were people who obviously both worked extremely hard and had minds that roamed and became somewhat obsessive. Like Jenner, Sir Humphry Davy even wrote poetry, which he published.[65] These were truly multitalented people who were hard to pigeonhole, despite coming down through history for famous inventions (e.g., the Davy lamp) and work on chemistry (e.g., nitrous oxide: laughing gas, isolation of several elements including sodium and potassium[66]). It is, I think, rarer in the 21st Century for someone to have such wide-ranging publications. Scientists are chasing metrics and assessment criteria, such as the Research Excellent Framework (REF) in the UK. Impact factors and citation indices as well as funding are driving science and I fear preventing such crazy investigations such as watching ants. As the climate of the Earth is now changing, I hope that the imagination and sheer fascination of scientists is not being quashed. Looking at nature, whether you are a physicist or an aeroengineer, or a poet, or a person simply walking down the street, can be inspirational,[67] and there is still a lot our animals can teach us.

6.8   Climate change

One of the tenets of this text is that animals can be used to assist in our understanding of climate change and how the environment may be altering in the future. It has been suggested that all areas of the world need early weather warning systems.[68] This is a big ask, and will be hugely time consuming and expensive. Perhaps there is a place here for animals and their behaviours to help in this?

As I write, the COP26 conference has concluded. Did it succeed? That depends on who is being quoted.[69] The intention was to try to get nearly 200 countries to commit to pledges on their use of fossil fuels so that the global temperature does not rise to more than 1.5 °C above that of pre-industrial periods. At the eleventh hour, after two weeks of negotiations the final wording was watered down so that coal use was going to be phased down rather than phased out and the prediction is that the global warming is more likely to level out at about a rise of 2 °C or a little more, relative to the pre-industrial level.[70] However, for some island states, such as those in the Pacific like the Marshall Islands, this has been a big failure[71] and they are seriously worried that their countries will be wiped from the face of the Earth, a little like Pompeii was in 79 AD, but on a much slower timescale. Only time will tell how the climate will change in the future, but there is almost certainly going to be continued loss of ice at the poles and on glaciers, and therefore rising sea levels, and there are likely to be more regular, and more severe, devasting events, such as flooding, drought, cyclones (typhoons and hurricanes) and even volcanic eruptions and earthquakes. There is a growing imperative to prevent this from happening, but also to try to predict severe events. Hopefully the robust study of how the organisms around us not only respond to, but also appear to predict, the environmental changes will be helpful as we enter a rather uncertain future. What is almost certain is that animals will adapt to climate change. Many will migrate to areas which are favourable for them, or move away from areas which are now not able to sustain their populations. However, some will be lost. The world will be different, and reversing the damage done may not be possible. Some regions will become uninhabitable for humans, and in some cases this may be a benefit to nature. When the Chernobyl nuclear power station exploded on 26th April 1986 near Pripyat, Ukraine, there was an exclusion zone set up. Humans were cast out. Nature was left to itself, and ironically the area is now a better place for nature than it was before. The radiation seems to have been tolerated by many of the plants and animals left behind and it is said to have become a haven for wildlife.[72] It is said that "nature abhors a vacuum",[73] (horror vacui), a saying attributed to Aristotle, but it seems that sometimes nature abhors humans. Without us, nature seems to thrive. If the human race became extinct, the world would not end, it would just be different. For many organisms it might be better. A similar situation occurred temporarily during the COVID-19 pandemic, caused by the spread of the SARS-CoV-2 virus. Many countries imposed lockdowns, restricting the movement of humans. Some of the animals thrived, with strange reports of puma in the streets of Santiago, Chile,[74] and groups of ducks crossing the roads of Paris, France.[75] It seemed as though the animals relished our removal.

One argument against the climate change data being used to persuade governments to alter their peoples' behaviour is that the climate has fluctuated since the formation of the Earth. Some argue that it is not anthropogenic activity that is the problem, but that climate change may be caused by normal changes of the sun's activity, or that it simply is not real.[76] However, many activities look at the climates in the past. These can be gleamed from the width of tree rings for example, as the better the season the more the tree grows and the wider the tree ring will appear.[77] This is referred to as dendrochronology. Interestingly, such studies can date violins and other stringed instruments,[78] and the data from such instruments can also be used to estimate the state of the climate when the trees were felled.[79] Alternatively, researchers use ice cores[80] and can then gleam what the climate was like when the ice was formed, often from the snow fall which gets compressed into ice. Graphs which are created of temperature over time do have a startling rise which starts around the industrial revolution[81] of the 18th Century and have been steadily rising ever since. Such data are hard to ignore, and these rises in the global temperatures will have to be felt by the animals living out in the environment. Humans seem to forget that animals cannot don that coat or turn the heating up.

Animals can help us to understand the past too. Using isotope analysis and the examination of fossilised midge heads, researchers at the University of Liverpool, UK, could determine that the climate in the Northern Lancashire area was not as stable as first thought, with two times of abrupt climate change identified: 9,000 years ago and approximately 8,000 years ago.[82] Such studies can also give hints to how animals may survive climate change, as their epigenetic[83] evidence may fossilise.[84] As pointed out earlier, many animals will adapt. Epigenetic changes, which may be transferred down the generations may be helping here.[85] Lamarck was thought to be wrong in the face of Darwinian evolution, but with the discovery of epigenetics Lamarck's ideas are having something of a resurgence.[86] This is certainly an area of genetics which needs more attention and hopefully will help us understand how animals may respond to, and survive, climate change.

6.9   A final look at the biology behind animal sensing

Can any conclusions be drawn from this romp through the literature? Can animals really be used to measure and predict the weather and environment around us?

To answer these questions requires knowledge of what the animals are sensing. As discussed earlier, there is a range of physical properties of the environment which may be able to be perceived.

Odour, or smell, is perceived by a range of animals and it is possible that animals are perceiving chemicals in the air which give them a guide to what might be happening around them. It has been suggested that birds smell their surrounding as an aid to navigation, for example. The smell of rain, petrichor, is caused by geosmin, and this may be sensed by a range of animals. It is produced by bacteria to ward off nematodes, and these are lowly animals on the evolutionary scale. There is no reason so suggest that more advanced animals cannot smelt it also – humans can.

The sense of touch has been suggested to be important. It is likely that animals such as elephants are feeling the ground beneath their feet, and perhaps are able to perceive vibrations long before a serious earthquake hits. Birds too may be sensitive, and there is no reason to rule out other animals too.

There are several other aspects of the environment that may be perceived. Temperature differences may be sensed, along with changes in barometric pressure. However, before a final review of some of the evidence it should be noted that it will not be one characteristic of the environment which is important here.

At a cellular level the principles of cell signalling, i.e. how messages are perceived and responded to by cells, are often taught in a simplified fashion.[87] For example, one might discuss how adrenaline (epinephrine) is sensed by a cell. This requires a receptor (what is referred to as a G protein coupled receptor (GPCR)), and this leads to a cellular signalling pathway involving a G protein, the production of a small diffusible molecule called cyclic adenosine monophosphate (cAMP) and several enzymes. It is quite a complex system, and there is no surprise that lecturers tend to simplify this down into digestible chunks. Unfortunately, no one told the cell. Being bombarded with adrenaline after the individual had a shock does not preclude the cell from perceiving a host of other things at the same time. The cells will be awash with a range of molecules to which they may need to respond, including other hormones, cytokines and environmentally derived molecules, such as toxins. The cell needs to manage all this at the same time, so perception and responses are hugely complicated. If the cell cannot manage this perception of its environment, it might not survive.

As alluded to earlier, cell signalling systems also need to turn off. The signalling by adrenaline (epinephrine) is a good example. One has a shock, and their cells, and indeed body, respond and react. We have a heightened metabolism, more ATP production, increased breathing, better muscle metabolism, all so we can have a short-term fight and flight response.[88] However, no organism needs to stay in this heightened state for long. The elephant feels a tremor, it runs, but once on safe ground it needs to stop and relax. It is does not need to keep being so alert, and therefore the adrenaline response will wind down. All the molecules which were created in a cell signalling response can be destroyed by the cell (or in the extracellular medium). cAMP is broken down to AMP by an enzyme called phosphodiesterase, for example. Ca^{2+}, which cannot be created or destroyed by cells, but are hugely important in cell signalling, are moved and 'hidden' (in organelles[89] for example), for another round of future signalling. Molecules which are structurally altered, such as proteins, are restored to their previous state. For example, kinase enzymes can add a phosphate group to proteins (phosphorylation), and then once the response has done what it needed to do, phosphatase enzymes can take the phosphate off again (dephosphorylation). Again, this protein is then ready for another round of signalling. Perhaps the organism will be shocked again in the near future, and another flight/flight response needed. There are numerous other examples of protein alteration in this way, such as oxidation, S-nitrosylation etc. There are hundreds of protein modifications in what are referred to post-translational modifications,[90] as they occur on fully created proteins, that is, after the process of protein translation when the amino acids have been joined together. Therefore, cells will respond to an environmental cue, the response will last for a short time, and then the cell 'relaxes' ready for future cues. And so it goes on. Biochemistry is one of the world's greatest recyclers. ATP/ADP, NAD^+/NADH, phosphorylation, inositol compounds, etc are all recycled. Cells do not always make new compounds from scratch. As mentioned earlier, most animals are in survival mode, and do not have resources to squander, so their cells recycle. How does this have relevance to our discussion? Animal behaviour changes are likely to be short-lived. A frighten animal will not remain frightened. The animal may know something is amiss, but if you come back the next day, it might look very relaxed. The behaviour change may well have come and gone, and no one noticed. Therefore, understanding behaviour and then creating a technology to mimic it may be a better way forward than continuing to observe the animals. Tracers on goats etc. may not be such a mad idea. The outputs can be continuous and auto-alerts sent, leaving the goats to get on with their lives. Alternatively, perhaps sensors can be produced which require no animal at all.

In the example with elephants given by Meenakshi and Juvanna[91] earlier, the animals ran, were brought back and then they cried. Clearly, they were spooked, and there would have been an adrenaline fright/flight response. The adrenaline receptors would have initiated the cell signalling pathways to increase metabolism and heightened muscle activity allowing them to run, fast into the trees and to higher ground. Had the elephants been left in the forest they would have been able to relax. Bringing them back was clearly traumatic and they could not relax. Instead, it was reported that they cried. These elephants were still in an alerted state, and their biochemical processes would not have been able to be transitioned back to the 'basal' or 'relaxed' state. Those elephants were still convinced that there was danger near, and their bodies, due to the amended biochemical processing, were still ready to react. And then the tsunami hit!

Very recently a crying response was also reported in dogs. In this case it was because of a positive event: they were happy to see their owners.[92] This was thought to be the first time such a response was documented. Such articles, however, do highlight that a crying response is not limited to humans, and at least other mammals are also capable of a crying-manifesting emotion, either to a stress, as in the elephants, or to happiness, as with the dogs.

Of relevance to this argument about animal behaviour is also the fact that we are not just machines that blindly follow instructions and environmental cues. We have sentience. We have an awareness of our surroundings. It may not be much of a stretch to suggest that gorillas are similar. We can imagine that an elephant is sentient, and therefore when it perceives the environment it will have an element of decision making too. If the ground shakes, does the elephant simply respond as a machine would, or does it make a conscious decision to run? It is not hard to see that an elephant may weigh up all its options and decide to act, usually with self-preservation most keenly on its mind. However, what of the ant? Is that animal simply a machine? Is it just reacting blinding to whatever it encounters? If we wander back down the evolutionary scale from humans to the sludge we originally emanated from, at which point on the animal kingdom would we expect to find sentience in animals? This is a question which is hard to answer. However, many are trying to get to grips with this issue. Partly this is driven by the desire not to cause unnecessary harm and distress to animals which we tend to use with little thought. In the UK there has recently been the reading of the Animal Welfare (Sentience) Act 22.[93] The definition of animal here is any vertebrate other than *Homo sapiens*, any cephalopod mollusc and any decapod crustacean, and the Act gives rights to these animals on the basis of their sentience.

But what of so-called lower animals? Some argue that sentience may be far wider than we first imagined. Helen Lambert, for example, an animal welfare research consultant and global sentience expert, has written quite extensively on the topic. She has looked at the evidence of sentience in reptiles,[94] amphibians,[95] fish,[96] and insects.[97] Of particular pertinence here is one of the "Highlights" in the paper on insects, which says that the authors: "Found evidence of cognitive capacities and sentience in a range of insect species." Very recently, Matilda Gibbons *et al.*[98] at Queen Mary, University of London, have been working on whether insects can feel pain, and there appears to be some evidence that they can. It cannot be assumed, therefore, that if insects are scurrying away that is simply a mechanical response to whatever they perceived. Perhaps such lower animals are making some decision based on a range of senses and they, like the elephants, are showing an informed behaviour which will optimise their survival. One would assume it would be the same for the arachnids. Very recently, eye movement in jumping spiders suggests that they enter a REM-like sleep state, and it has been posited that they might dream.[99]

However, there is some argument against sentience being important in insects' behavioural changes observed. In Harlow Shapley's original paper, he posits that the ants were slowing down because they were content to have some respite from the heat. This suggests that Shapley was assuming some sentience by the ants, which were responding in a way which made it appear that they had some awareness of their environment. However, the data suggest that the ants were simply responding to the temperature in a mechanistic manner, and their run speed showed no hint of the influence of sentience. However, the same argument could be used for the rattlesnakes, and these are now thought to be more likely to show sentience. Clearly there is more understanding needed here. Of course, Helen Lambert and her team are not the only ones to be investigating sentience in lower animals[100] and I am sure much more will be written about sentience in a range of lower animals in the future.

On a larger scale than the cell, that is, the whole animal, the same principles of perception of the environment are relevant. Animals will need to perceive and respond to more than one characteristic of the environment at the same time. An elephant will not stop seeing just because it hears things. Both senses will work simultaneously, and those perceptions will be unscrambled in the brain so that the elephant as a whole can respond. Therefore, the change in the environment which may indicate impending disaster may be perceived by the elephant as sound (perhaps outside the range we can sense), sight, touch (as in vibrations), smell or even environmental factors we have yet to fully understand, such as the sensing of air pressure. Once all these signals are processed simultaneously the elephant may decide to run for the trees. And of course, it might just run for the trees because other elephants have decided to leave, and the individual elephant is just spooked by its community group. Or it may have seen other animals scarper. Whatever the reason, if humans see the elephants leave, and know that this is not a good thing, they too can take a cue from their observations and brace for whatever may be coming. In many ways it does not matter why the elephant responds (although scientifically interesting), as long as the observation is valid to indicate changes in weather or an earthquake, and then it can be used by humans as a forecast. It needs to be remembered that many of these observations have been known about, and written about, for hundreds of years, with little scientific underpinning.

So, let's revisit some of the evidence of what animals may be sensing.

There seems little doubt that there is a range of cold-blooded animals that can be used as a proxy thermometer. This is summed up by Dolbear's Law, although it transpired that he was not the first to notice this, or indeed publish on it. For that we need to thank Margarette W. Brooks, who pointed out that a W.G.B. had already written about it. This was years before Dolbear. It is tempting to wonder if W.G.B. was a woman, but there is no evidence for this, as the identity of this person has been lost through history. However, why else use their initials? Were they a woman writing in an age dominated by men? Regardless of who they were, the chirping of crickets can be used to estimate the temperature. This is true for the rattle of the tail of the rattlesnake, although the speed of vibration here would not make it easy to count without complex instrumentation. The running of an ant can also be used and is relatively easy to assess. Shapley claimed that one observation of ants can give the temperature with the accuracy of 1 °C, and that the animals react faster than his mercury thermometer.[101] However, the animal behaviour relationship with temperature is over a rather limited range. Cold-blooded animals are not happy at low temperatures, as found when snakes were placed in crushed ice.[102] This would give a lower limit somewhere above 4 °C. At the other end of the temperature scale approximately 40 °C was the upper limit. The central section of the graph between these limits was relatively linear, and could be used to estimate the air temperature. Therefore, theoretically an animal could be used as a thermometer. Interestingly, during the heat of the 2022 summer the cicadas in Provence stopped 'singing' when the temperature reached 36 °C and it is thought that the animals may move to cooler climates in the future.[103] Clearly, these animals have an upper temperature limit to their normal activity and other species would be the same.

There are many places on Earth that are below 4 °C, and many now that are regularly above 50 °C. When discussing climate change it is often these extremes which are discussed – Are the ice caps melting? Are glaciers sliding off mountains? Can the deserts still be inhibitable? *Newsnight*, a weekday TV news magazine on BBC2, has recently been running a series entitled *Life at 50 °C*.[104] This is because large areas of the world are experiencing this sort of temperature. On the 16th August 2020, the temperature at Furnace Creek in Death Valley was recorded as 54.4 °C (129.9 °F).[105] When it needs to be considered that protein denaturation may occur above approximately 41 °C,[106] and certainly will happen for most proteins in the mid-50s °C, then such temperatures are not very compatible with using animals as thermometers at these extreme temperatures now being seen more commonly around the globe. Using animals as a way to measure air temperature is likely to remain an interesting observation, and never be used in a pragmatic manner. On the other hand, understanding why, and how, temperature alters the activity of cells and rates of metabolism is useful to unravel what is happening in many disease states and does inform biochemical endeavours. The use of proteins (particularly enzymes) in industry is also important, and increasing the temperatures at which enzymes can be used is critical, as anyone who uses biological washing powders appreciates, as a hot wash is incompatible, at least at the moment. Some enzymes do work at high temperatures, such as the polymerase from the thermophile *Thermus aquaticus*. This enzyme, referred to as Taq polymerase, is instrumental in the process of polymerase chain reaction (PCR), so instrumental in our fight against COVID-19. But for cold-blooded animals which are positing as a thermometer, such temperatures are simply not possible.

Outside of temperature measurements, the forecasting of the weather requires the observation of a range of other environmental conditions, including wind speed and direction, and barometric pressure. As can be seen from numerous examples earlier, many cultures, including in the UK, use the behaviour of animals to predict the forthcoming weather, either in the short- or mid-term. It almost seems like no animals are exempt from this. From insect and arachnids to birds and mammals, they all seem to get in on the act. Peoples around the world observe their nesting, their migration or simply strange behaviours, and then say what the weather will do. "A seal making its loud sound in the harbor while holding an octopus is a sign of storm."[107] Really? How often would this happen? However, such signs to those who believe it may be an indication of imminent rain, dry weather approaching, a storm or snow. I am sure that there is no basis for many of these myths and anecdotes. However, they are not new but span back through the generations, so clearly there is a belief in such observations.

There are however at least two things that regularly seem to underpin many of these observations. Firstly, the behaviour of an animal may be influenced by one not being observed. Birds may follow insects as a food source, perhaps soaring high into the sky. It is the behaviour of the birds which is being witnessed, and the insects are not considered, but they are the underpinning cause. These small, unseen animals may be being carried by air currents caused by changing weather, and the birds are merely feeding in an opportunistic way. So, the change of the action of the birds may actually make sense, and be a real indicator of the environmental conditions changing.

Secondly, there seems to be a body of evidence that animals can sense barometric pressure. As the weather around the world is dominated by cyclonic and anticyclonic regions, the air pressure is a major factor in causing weather. It is also a major way of predicting the weather to come. Here in the UK, we know that the arrival of a low-pressure region will bring a cold front and usually heavy rain, with a warm front following with more rain. On the other hand, a high pressure will bring dry weather, and in the winter cold, but warm in the summer. It seems that many animals can sense the changing air pressure and know that the weather is going to change, and they then alter their behaviour accordingly. These changes in the way the animals act, either in their movement to new places (into forests, or onto ceilings), or their new activity (building up nest defences), have been embedded into many cultures around the world, and are used by societies in many places to indicate environmental changes, perhaps as part of a broader indigenous knowledge system. But the fact that animals can sense barometric pressure shows that this is not just myth or hearsay, but is underpinned by a scientific basis.

Other predictions are also possible using animals. This includes earthquakes,[108] volcanos and tsunami. Of course, all of these natural phenomena are interrelated, and so it is no major surprise that if animals can be used as a predictor for one then they can be used for other such events too.

In a comprehensive review of earthquake prediction, Woith et al.[109] reviewed 180 articles where abnormal animal behaviour was reported. The data covered 160 earthquakes and encompassed the use of 130 species of animals. What were described as "precursor events" with animals were reported across 25 countries, but many were from the Pacific region (Japan, Taiwan, New Zealand), but also Europe (Italy). Interestingly, the reports of animal behaviour were slightly most frequent in February, April and May.

More importantly for the arguments here, were the animal behaviours reported before the earthquake? Many unusual behaviours were reported in the day before the earthquake struck, but the greatest level of observations was made in the hour preceding the quake, with the majority in the 5 minutes before. The lowest magnitude of earthquake reported to be predicted using animals was down to 2, but the data were better for more forceful earthquakes, i.e., above magnitude 6. And of no great surprise, the nearer the epicentre the more reports there were, with the majority being within 50 km.

Overall, the paper by Woith *et al.*[110] is quite critical of the data, and they suggest many ways in which the reporting of animal behaviour can be done in the future to improve the area of science. They point out that there are many factors not taken into account and that many reports appear to be retrospective, which is not very helpful if predictions of possible impending doom are going to be made. However, with a such a body of evidence it does show that animals do seem to have unusual behaviours before earthquakes and therefore careful observations can be a useful way to predict when they might happen. The warning may not be a long period in advance, and a day or less would not certainly be not enough to evacuate a major city, but it might be just enough to enable people to find open ground, for public transport to be stopped and for fuel supplies to be turned off.

Volcanoes and tsunami are also able to be anticipated by some animals. Both can cause major devastation and are going to happen in the future, perhaps with more frequency as climate changes happen and the global temperatures increase.

So, what is it that the animals are sensing? With the tremors emanating from the epicentre as P and S waves, travelling at different speeds, it may be that animals are sensitive enough to feel this and get a sense of something happening. This may be what elephants are sensing before they head inland. Alternatively, researchers talk about infrasonic pulses which are being detected by some animals, such as gulls. Infrasound is known to be produced by seismic activity,[111] so it is conceivable that animals may be able to sense this, as has been suggested.[112] It is also suggested that technological sensing of infrasound might be useful to predict natural events, such as tornadoes.

Animals may also be able to detect changes in electric potential or changes in magnetic forces. Proteins and protein interacting minerals[113] have been described and it is conceivable that such sensing systems are giving some animals an early warning system. This may be different for land and aquatic animals, but both are amongst the species listed as being observed as having abnormal behaviour.

The perception of magnetic fields by animals is not a particularly new idea. For example, James L. Gould looked at the possibility back in 1984,[114] reviewing the evidence that there was something to sense and whether animals could really perceive such changes. Others have suggested that magnetic deviations may account for unusual animal behaviour, such as the beaching of whales.[115]

One of the arguments that animals cannot sense impending events such as earthquakes is that there is no driver for the animals to evolve to do this. Earthquakes are not that regular or frequent if you are a relatively sessile animal, such as an insect which may not migrate too far from the nest. Therefore, why should an animal be affected? And few animals are likely to live long enough to remember the last event. However, this has been argued against, as animals have had billions of years of evolution.[116] It should also be remembered that as organisms have evolved, they may have developed a sensory system for one reason, but then it has, almost coincidentally, been useful for another reason. One of the big revelations of modern biochemistry is that many, perhaps the majority of, proteins have more than one function, and are referred to as moonlighting proteins.[117] These proteins almost certain evolved for one reason and then were found to be useful for a second, or even third, function in cells. A classic example here is that of the protein called cytochrome c. Cytochromes, which are proteins which contain a colour, were first described by Keilin in 1925.[118,119] Cytochrome c was found to be instrumental in the movement of electrons in the organelles called mitochondria[120] in cells.[121] In humans it is a relatively small and simple protein with a haem prosthetic group,[122] as found in haemoglobin. So, in short, it is hugely important in allowing our cells to have enough energy to carry out their functions. However, it was later found that the cytochrome leaves the mitochondria and triggers the cells to die,[123] in a process dubbed as apoptosis.[124] This is cell suicide. The cells are forced to die, and one of the mechanisms is that cytochrome c drives this to happen. Therefore, this protein has two very disparate functions. In the control of cell function some of the components used are extremely toxic and it appears that cells have not only evolved to survive their presence, but they have adopted them for positive use. This includes such noxious gases as are released from volcanoes, such as hydrogen sulfide[125] (presently, an Italian island called Volcano, north of Sicily, is being partially evaluated for this reason[126]). In this context it is not a great leap to suggest that animals have adopted sensory systems which may have been evolved to aid in communication, mating and prey/predation to sense the environment as a way of escaping natural disasters. Afterall, it is better for an animal to think an earthquake is coming and act, and then be wrong, than doing nothing at all and being caught when the tremors hit. So, the anticipatory system does not have to be very accurate, as long as it is sensitive enough. Of course, if that is the case it makes our use of abnormal activity hard to use as a true predictor, and perhaps this has been one of the problems through history. There are many reasons why animals might be showing unusual behaviour, and these must have been noted regularly, without any impending doom for the location in which they live. They might be tired, hungry, ill, injured or just depressed. That does not mean that there is a typhoon around the corner.

Of course, there will remain a significant number of myths and anecdotes which will not be explained by science. These will remain as "old wives' tales".[127] In her Master of Arts thesis in 1976, entitled *The Reliability of Selected Weather Beliefs*, Judith K. Sadewasser[128] separates the weather indicators into two sections. The first is "Analysis of twenty weather beliefs having a scientific validity". This includes such things as "Red sky at night", which is supposed to be shepherds' delight, as in the weather the next day will be favourable. Others included here are: "When smoke goes to the ground", which is because the air is humid and rain likely, and "A person's hair gets curly", which again is a sign of rain. Hair can absorb water, it is hydroscopic, and therefore humidity can affect hair and be a sign of rain approaching – as used in some hygrometers and discussed earlier. In the second section, Sadewasser has a list of what she describes as "A sampling of weather beliefs having no scientific validity". Here are several which involved animals, including "If a rabbit's fur is thick" (a sign of a harsh winter ahead), "When a crawfish's chimney is plugged" (a sign of rain, but seems to be contrary to the actual behaviour of the animal), "If a groundhog sees his shadow, there will be 40 days of bad weather" (which seems to be an altered myth from the time Germans emigrated to North America) and "If you step on a toad, it's going to rain". However, of particular relevance to the earlier discussion the woolly worm is in this section too: "If the woolly worms have a wide ring around then it's gonna be a bad winter." Sadewasser argues that the banding of the caterpillar is an indicator of weather past, not weather future, as discussed in Chapter 3.

Many of the indicators of weather that use animals will remain unexplained, and thought to be bizarre, but if people find they give some hint of what might be coming then they will no doubt remain embedded in cultures for generations to come, just as they have sustained in the generations through history. However, some are simply ridiculous. In Ewer,[129] he finishes his paper with this statement:

> Finally, the reader may be cautioned that should he find a dragon's egg in a lake or river, the sage Yuin-Chu-Tsih tells us that it means that there will certainly be a flood.

Of course, dragons don't actually exist, despite what you may see or read in the Harry Potter works,[130] and so the finding of a dragon's egg is not likely. The only dragon on Earth, which is a very dangerous and fierce lizard, is the Komodo dragon (*Varanus komodoensis*),[131] and although they can reach 10 feet in length, they do not hiss fire. There are of course many other animals with the word "dragon" in their name, including the Blue Dragon Sea Slug, Pink Dragon Millipede and Black Dragonfish.[132] True dragons, on the other hand, do not exist, so finding an egg from one will never happen, despite what the sage tells us.

Throughout the preceding discussion there have been sections on the technologies which are available and how these have been developed – summarised on Table 3. It is quite ironic that the first earthquake detector was based on balls falling from dragons into the mouths of frogs.[133] Can the use of animals come around full circle? Can we now start to design instruments to predict the weather and natural phenomena such as earthquakes based on what the animals are sensing? If it can be determined what the animals are sensing, be that P waves, or changes in geomagnetism, can we develop sensors that can do the same? The work of Michael Garstang and Michael C. Kelley,[134] for example, where they suggest that changes

A Komodo dragon (*Varanus komodoensis*).
(Sourced from Shutterstock. Contributed by BeautifulBlossoms. Image made black and white.)

in the electrical fields and radio waves are converted to sound which are sensed by animals, could well become the basis of a technological solution to the prediction of earthquakes and tsunami. The speed (speed of light: 186,000 miles/second[135] or 300,000 km/second) of the radio waves would outpace the sounds waves from the movement of the Earth's crust (speed of sound at sea level: 0.21 miles/second,[136] or 0.34 km/second) and give an early warning. The warning would be greater the farther from the epicentre of a quake, but it may still be long enough to mitigate potential damage and fatalities. Such technologies could form a remote sensing system which can sound alarms, rather than relying on the observation of animals, which is time consuming and not particularly robust. Perhaps a cow will simply sit down as it is tired, and then others copy. It does not mean rain is on the way. But if they are sitting down because of a suite of senses, including barometric pressure (which of course we measure anyway), then perhaps it does indicate rain, and we can use the same sensing in a more sensitive and remote instrument. This principle may be extended to extreme events, such as volcanic eruptions, earthquakes and resulting tsunami. Perhaps it is time to have a better, and less sceptical, look at the unusual behaviour of the animals around us.

6.10   Conclusions and future

Having surveyed the literature on the use of animals to sense and predict our environment let me return to the horses on the hills above St Austell Bay in 1999. Did they sense the arrival of the eclipse well before I did (although of course I knew it was coming by reading about it for weeks, not an option for the horses)? Perhaps they did? Perhaps they could sense the change in air pressure as the cooler air was approaching? Perhaps their perception of their environment is much better than mine? Certainly, equines have been discussed earlier, including in the prediction of earthquakes.[137] Maybe it was not just my imagination that led to observation of the unusual behaviour of our equine friends? Perhaps they were trying to tell us something after all? Although video evidence from Idaho would suggest otherwise.[138]

Perhaps humans are becoming more and more disconnected from nature and what is around us, as has been suggested by some.[139] Here, a question is posed and then answered:

It is widely accepted that we are more disconnected from nature today than we were a century ago, but is that actually true? A recent study we conducted suggests that it is – and that may be bad news not only for our well-being but also for the environment.

With climate change now being one of the major issues of the early 21st Century,[140] it might be time to reverse this trend and reconnect with nature. Observing the animals, and plants, should be a large part of this.

But what of us humans? Often it is forgotten, or rather overlooked, that humans are animals too. Therefore, if all the species listed earlier can measure or sense the climate, can we? As already argued in Chapter 2, we are warm-blooded, so we will not run about faster as the temperature rises as we see with Shapley's ants.[141] However, recent evidence suggests that humans can sense the lunar cycle.[142] Does that mean we are also sensitive to geomagnetic forces? Does that mean we too can sense the imminent earthquake? A quote which can be found on the internet is:

> Neither the USGS nor any other scientists have ever predicted a major earthquake. We do not know how, and we do not expect to know how any time in the foreseeable future.[143]

[Listed as being posted 15th January 2020]

Therefore, as this is an all-encompassing statement, it includes the human use of technology as well as human senses. As an animal, if we were ever capable of sensing the changes that may prelude an earthquake, we seemed to have forgotten it. On the other hand, when people say that their joints ache and therefore bad weather is on the way, maybe they are being very astute. There is certainly evidence that pain can be altered by factors such as barometric pressure and this may be related to the onset of a low-pressure atmospheric region, and hence rain or storms. As animals, we may not be immune to acting as weather sensors.[144]

The whole topic of this book remains somewhat contentious, not helped by my use of mythical animals such as dragons! There are dozens of myths and anecdotes about animals (and plants) and weather forecasting, or predicting the impending doom of an earthquake, and some are taking some of this science seriously, setting up their zoo seismic centres.[145] It is no doubt a subject that will persist into the future. As more is uncovered about how animals sense, it may be that some of the anecdotes may be taken more seriously, and as mentioned, perhaps new technologies will emerge based on new findings.

It is fitting, however, that this book should have a final quote (albeit quite a long one) from one of my Cornish heroes, Sir Humphry Davy. Although born in Penzance, Cornwall, he died overseas, on 29th May 1829, in Geneva, Switzerland. Before he died, he wrote a work which is referred to as *Salmonia*,[146] but it actually goes by a very long title of *Salmonia: or Days of Fly Fishing (in conversations with some account of the habits of fishes belonging to the genus Salmo by an angler).* The final work was edited by his brother, John Davy. The book is based on the exploits of four people who go fishing and whilst there they start discussions on a range of topics, including the weather. The four characters are fictitious and are named Halieus, Poietes, Ornither and Physicus. Despite being written approximately 200 years ago, Davy was pondering the same issues as discussed in this book. A pertinent section of this conversation is:

> POIET.[147]- I have often seen sea-gulls assemble on the land, and have almost always observed, that very stormy and rainy weather was approaching. I conclude that these animals, sensible of a current of air approaching from the ocean, retire to the land to shelter themselves from the storm

However, even Davy argues against himself here:

> ORN.[148]- No such thing. The storm is their element; and the little petrel enjoys the heaviest gale, because, living on the smaller sea insects, he is sure to find his food in the spray of a heavy wave – and you may see him flitting above the edge of the highest surge. I believe that the reason of this migration of sea-gulls, and other sea birds, to the land, is their security of finding food. They may be observed, at this time, feeding greedily on the earth worms and larvae, driven out of the ground by severe floods; and the fish, on which they prey in fine weather in the sea, leave the surface, when storms prevail, and go deeper. The search after food, as we agreed on a former occasion, is the principal why animals change their places. The different tribes of the wading birds always migrate, when rain is about to take place; and I remember once, in Italy, having been long waiting, in the end of March, for the arrival of the double snipe in the Campagna of Rome, – a great flight appeared on the 3rd of April, and the day after, heavy rain set in, which greatly interfered with my sport. The vulture, upon the same principle follows armies; and I have no doubt that the augury of the ancients, was a good deal founded upon the observation of the instincts of birds. There are many superstitions of the vulgar, owing to the same source. For anglers, in spring, if is always unlucky to see single magpies, – but *two* may always regarded as a favourable omen; and the reason is, that in cold and stormy weather, one magpie alone leaves the nest in search for food, the other remaining sitting upon the eggs, or the young ones; but when two go out together the weather is warm and mild, and thus favourable for fishing.

In the original draft of this book, I concluded with the statement: "Enough said, I feel." However, a couple of further points ought to be added. Firstly, if you still remain sceptical having read this far, I hope this book has at least made you think. Clearly some of the reports regarding the use of animals are never going to be useful, but hopefully a healthy level of scepticism has shown you that some of the information may be valuable. Lastly, with weather being ever important, and with climate change making a tangible impact, hopefully observations of animals and their behaviour will indicate how the world is changing and how we can live with those changes. Perhaps, in the end, observing our environment needs to be taken more seriously, and the animals may be telling us something, even if we do not always understand them.

Endnotes

1 Long, J.A. and Stoy, P.C. (2014) Peak tornado activity is occurring earlier in the heart of "Tornado Alley". *Geophysical Research Letters*, 41, 6259–6264.
2 An example of a tornado chasing company running tours. Extreme Tornado Tours. com: https://extremetornadotours.com/ (Accessed 08/08/22)
3 McGuire, B. (2022) *Hothouse Earth: An Inhabitant's Guide*. Icon Books, London. ISBN: 978-1785789205. Bill McGuire is Professor Emeritus of Geophysical and Climate Hazards at University College London. His is also a co-director of the New Weather Institute and contributed to the 2012 IPCC report.
4 Dr Stuart Parkinson is Executive Director of *Scientists for Global Responsibility* (www.sgr.org.uk/ (Accessed 03/08/22))
5 The Guardian: www.theguardian.com/environment/2022/feb/13/early-spring-changing-behaviour-flora-fauna-climate-change (Accessed 13/02/22)
6 The Meteorological Office (UK): www.metoffice.gov.uk/about-us/press-office/news/weather-and-climate/2022/red-extreme-heat-warning-ud (Accessed 03/08/22)
7 The Meteorological Office (UK): www.metoffice.gov.uk/about-us/press-office/news/weather-and-climate/2022/driest-july-in-england-since-1935 (Accessed 03/08/22)
8 The BBC: www.bbc.com/future/article/20151214-why-do-brits-talk-about-the-weather-so-much (Accessed 02/11/21)
9 Enock, C.M. (2013) Indigenous knowledge systems and modern weather forecasting: exploring the linkages. *Journal of Agriculture and Sustainability*, 2(2).
10 "Uncle Offa" (1991) *Natural Weather Wisdom*. The Self Publishing Association Ltd., Worcs, UK.
11 A fictious walking plant, which was a result of an accident. From: *The Day of the Triffids* by John Wyndham, Penguin. London. ISBN: 9780141033006
12 Corlett, R.T. and Westcott, D.A. (2013) Will plant movements keep up with climate change? *Trends in Ecology & Evolution*, 28, 482–488.
13 Viana, D.S. (2017) Can aquatic plants keep pace with climate change? *Frontiers in Plant Science*, 8, 1906.
14 The Guardian: www.theguardian.com/environment/2022/jan/13/plants-at-risk-of-extinction-as-climate-crisis-disrupts-animal-migration (Accessed 25/01/22)

15 Moura, G.S. and Franzener, G. (2017) Biodiversity of nematodes biological indicators of soil quality in the agroecosystems. *Arquivos do Instituto Biológico*, 84. [This is in English]

16 Ruess, L., Michelsen, A., Schmidt, I.K. and Jonasson, S. (1999) Simulated climate change affecting microorganisms, nematode density and biodiversity in subarctic soils. *Plant and Soil*, 212(1), 63–73.

17 Ackerman, J. (2016) *The Genius of Birds* (p. 230). Corsair, London. ISBN: 9781472114365

18 Eira, I.M.G., Oskal, A., Hanssen-Bauer, I. and Mathiesen, S.D. (2018) Snow cover and the loss of traditional indigenous knowledge. *Nature Climate Change*, 8(11), 928–931.

19 Cahill, A.E., Aiello-Lammens, M.E., Fisher-Reid, M.C., Hua, X., Karanewsky, C.J., Yeong Ryu, H., Sbeglia, G.C., Spagnolo, F., Waldron, J.B., Warsi, O. and Wiens, J.J., 2013. How does climate change cause extinction? *Proceedings of the Royal Society B: Biological Sciences*, 280(1750), 20121890.

20 A genome is the complete complement of DNA in a species. DNA amplification can be carried out using the polymerase chain reaction (PCR), as has become commonplace during the COVID-19 pandemic to test for people being viral positive.

21 Shapiro, B. (2015) *How to Clone a Mammoth: The Science of De-Extinction*. Princeton University Press, Princeton. ISBN: 9780691173115

22 The characters in the book tried to establish an ecosystem of dinosaurs on an island, but it was not sustainable. The destructive and predatory ones escaped their enclosures and wreaked havoc, which was not good for the people visiting.

23 The film was based on a book by Michael Crichton: *Jurassic Park*. Arrow, London. ISBN: 9781784752224

24 Manyanhaire, I.O. (2015) Integrating indigenous knowledge systems into climate change interpretation: Perspectives relevant to Zimbabwe. *Greener Journal of Educational Research*, 5, 27–36.

25 Enock (2013).

26 Green, D., Billy, J. and Tapim, A. (2010) Indigenous Australians' knowledge of weather and climate. *Climatic Change*, 100(2), 337–354.

27 Elia, E.F., Mutula, S. and Stilwell, C. (2014) Indigenous Knowledge use in seasonal weather forecasting in Tanzania: the case of semi-arid central Tanzania. *South African Journal of Libraries and Information Science*, 80(1), 18–27.

28 Page, S. (2002) *Astrology in Medieval Manuscripts*. University of Toronto Press, Toronto, Canada.

29 Owens, D. (2019) *Where the Crawdads Sing*. Corsair, London. ISBN: 978-1472154668

30 A Crawdad is a freshwater crayfish.

31 The University of Georgia: www.ecology.uga.edu/inspired-by-nature-delia-owens/ (Accessed 03/08/22)

32 IMDb: www.imdb.com/title/tt9411972/ (Accessed 03/08/22)

33 Lantz, T.C. and Turner, N.J. (2003) Traditional phenological knowledge of Aboriginal peoples in British Columbia. *Journal of Ethnobiology*, 23(2), 263–286.

34 The shad bush appears to be even named after the fish, as there is such a cultural interrelationship between the two. It is also called the serviceberry as it blooms when the ground has thawed sufficiently for burial services. NYBG: www.nybg.org/blogs/plant-talk/2012/04/learning/native-plants-101-the-shadbush-story/ (Accessed 06/01/22)

35 Almanac: www.almanac.com/predicting-weather-plants (Accessed 06/01/22)

36 Missouri Department of Conservation: https://mdc.mo.gov/discover-nature/field-guide/scarlet-pimpernel-poor-mans-weatherglass (Accessed 06/02/22)

37 The Natural Navigator: www.naturalnavigator.com/the-library/weather-lore/ (Accessed 06/02/22)

38 "Uncle Offa" (1991).

39 NIWA: https://niwa.co.nz/sites/niwa.co.nz/files/Traditional-Maori-Weather-and-Climate-Forecasting-poster.pdf (Accessed 06/01/22)

40 It is worth pointing out that this site also uses celestial markers for weather forecasting too. For example, if the moon is lying on its back a good month is coming.

41 Nedelcheva, A. and Dogan, Y. (2011) Usage of plants for weather and climate forecasting in Bulgarian folk traditions. *Indian Journal of Traditional Knowledge,* 10(1), 91–95.

42 World Animal Protection: www.worldanimalprotection.org.uk/blogs/concerned-about-climate-change-eating-less-meat-reduces-greenhouse-gas-emissions?gclid=Cj0KCQiAw9qOBhC-ARIsAG-rdn43TTxaqe1c6DObbqWkJdxqEXyFxS8prUR_4f-smmOS4acakbDudqgaAvPWEALw_wcB (Accessed 06/01/22)

43 Almanac: www.almanac.com/predicting-weather-plants (Accessed 06/01/22)

44 Brandywine Conservancy: www.brandywine.org/conservancy/blog/rhododendrons-thermometers (Accessed 06/02/22)

45 Wang, X., Arora, R., Horner, H.T. and Krebs, S.L. (2008) Structural adaptations in overwintering leaves of thermonastic and nonthermonastic Rhododendron species. *Journal of the American Society for Horticultural Science*, 133(6), 768–776.

46 Royal Botanic Gardens Kew: www.kew.org/read-and-watch/victorian-plant-hunters-china (Accessed 06/01/22)

47 Britannica: www.britannica.com/science/sentinel-species (Accessed 03/07/22)

48 Beyer, J., Green, N.W., Brooks, S., Allan, I.J., Ruus, A., Gomes, T., Bråte, I.L.N. and Schøyen, M. (2017) Blue mussels (*Mytilus edulis* spp.) as sentinel organisms in coastal pollution monitoring: A review. *Marine Environmental Research*, 130, 338–365.

49 Sedeño-Díaz, J.E. and López-López, E. (2013) Fresh water fish as sentinel organism: From the molecular to the population level, a review. *New Advances and Contributions to Fish Biology*. In H. Türker (ed.), *New Advances and Contributions to Fish Biology*. IntechOpen, London.

50 Carbone, D. and Faggio, C. (2019) *Helix aspersa* as sentinel of development damage for biomonitoring purpose: A validation study. *Molecular Reproduction and Development*, 86(10), 1283–1291.

51 Baroudi, F., Al Alam, J., Fajloun, Z. and Millet, M. (2020) Snail as sentinel organism for monitoring the environmental pollution; a review. *Ecological Indicators*, 113, 106240.

52 Sures, B. (2001) The use of fish parasites as bioindicators of heavy metals in aquatic ecosystems: A review. *Aquatic Ecology*, 35(2), 245–255.

53 Dahiya, U.R., Das, J. and Bano, S. (2022) Biological indicators of soil health and biomonitoring. In *Advances in Bioremediation and Phytoremediation for Sustainable Soil Management* (pp. 327–347). Springer, Cham.

54 Adrian, R., O'Reilly, C.M., Zagarese, H., Baines, S.B., Hessen, D.O., Keller, W., Livingstone, D.M., Sommaruga, R., Straile, D., Van Donk, E. and Weyhenmeyer, G.A. (2009) Lakes as sentinels of climate change. *Limnology and Oceanography*, 54(6part2), 2283–2297.

55 Schindler, D.W. (2009) Lakes as sentinels and integrators for the effects of climate change on watersheds, airsheds, and landscapes. *Limnology and Oceanography*, 54(6part2), 2349–2358.

56 Bossart, G.D. (2011) Marine mammals as sentinel species for oceans and human health. *Veterinary Pathology*, 48(3), 676–690.

57 Wormworth, J., Sekercioglu, C.H. and Şekercioğlu, C. (2011) *Winged Sentinels: Birds and Climate Change*. Cambridge University Press, Cambridge, UK.

58 Ferreira, F., Santos, M.M., Castro, L.F.C., Reis-Henriques, M.A., Lima, D., Vieira, M.N. and Monteiro, N.M. (2009) Vitellogenin gene expression in the intertidal blenny *Lipophrys pholis*: A new sentinel species for estrogenic chemical pollution monitoring in the European Atlantic coast? *Comparative Biochemistry and Physiology Part C: Toxicology & Pharmacology*, 149(1), 58–64.

59 Wallisch, K. (1999) *Animal Behavior as a Weather Predictor* (Doctoral dissertation, Oklahoma State University). Although Google Scholar cites this as a doctoral thesis, it was in fact submitted for a Master of Science.

60 Ewer, D.W. (1952) Animal weather prophets. *Weather*, 7, 16–19.

61 Quotes of Famous People: https://quotepark.com/quotes/1729350-jonathan-swift-tis-very-warm-weather-when-ones-in-bed/ (Accessed 06/01/22)

62 Poetry Foundation: www.poetryfoundation.org/poets/jonathan-swift (Accessed 06/01/22)

63 Poetry Foundation: www.poetryfoundation.org/poets/john-clare (Accessed 06/02/22)

64 "Uncle Offa" (1991).

65 Holmes, R. (2008) *The Age of Wonder: How the Romantic Generation Discovered the Beauty and Terror of Science*. Harper Press, Toronto, Canada. ISBN: 9780007149537

66 University of Waterloo: https://uwaterloo.ca/chemistry/community-outreach/2019-international-year-periodic-table-timeline-elements/davys-elements-1805–1824 (Accessed 30/12/21)

67 Royal Aeronautical Society: www.aerosociety.com/news/engineering-nature/ (Accessed 12/12/21)

68 The Guardian: www.theguardian.com/world/2022/mar/23/un-chief-calls-for-extreme-weather-warning-systems-for-everyone-on-earth (Accessed 23/03/22)

69 The Independent: www.independent.co.uk/climate-change/infact/cop26-glasgow-climate-agreement-success-failure-b1957446.html (Accessed 15/11/21)

70 PLOS: https://allmodels.plos.org/what-is-success-cop26-and-beyond/ (Accessed 15/11/21)

71 The Guardian: www.theguardian.com/world/2021/nov/15/cop26-pacific-delegates-condemn-monumental-failure-that-leaves-islands-in-peril (Accessed 15/11/21)

72 UN Environment Programme: www.unep.org/news-and-stories/story/how-chernobyl-has-become-unexpected-haven-wildlife (Accessed 12/12/21)

73 Grant, E. (1973) Medieval explanations and interpretations of the dictum that 'Nature Abhors a Vacuum'. *Traditio*, 29, 327–355.

74 France 24: www.france24.com/en/20200403-pumas-in-santiago-wildlife-takes-to-cities-amid-coronavirus-lockdown (Accessed 05/08/22)

75 Reuters: www.reuters.com/article/earth-day-france-birds-idUSL5N2CA1Y4 (Accessed 05/08/22)

76 Help Save Nature: https://helpsavenature.com/arguments-against-global-warming (Accessed 02/11/21)

77 Trouet, V. (2020) *Tree Story: The History of the World Written in Rings*. John Hopkins University Press, Baltimore, USA. ISBN: 9781421443744

78 Čufar, K., Beuting, M., Demšar, B. and Merela, M. (2017) Dating of violins–the interpretation of dendrochronological reports. *Journal of Cultural Heritage*, 27, S44–S54.

79 Wilson, R. and Topham, J. (2004) Violins and climate. *Theoretical and Applied Climatology*, 77(1), 9–24.

80 Brook, E.J. and Buizert, C. (2018) Antarctic and global climate history viewed from ice cores. *Nature*, 558(7709), 200–208.

81 Ashton, T.S. (1997) *The Industrial Revolution 1760–1830*. OUP Catalogue, Oxford University Press, Oxford, UK. ISBN: 9780192892898

82 Science Daily: www.sciencedaily.com/releases/2007/07/070709111430.htm (Accessed 02/11/21)

83 Epigenetics: changes to the DNA of an organism which does not alter the base sequences. For more see: Carey, N. (2012) *The Epigenetics Revolution: How Modern Biology Is Rewriting Our Understanding of Genetics, Disease, and Inheritance*. Columbia University Press, New York. ISBN: 9781848313477

84 Holmes, B. (2012) Clues to surviving rapid climate change in fossil DNA. *New Scientist*, 213, 8–9.

85 Bošković, A. and Rando, O.J. (2018) Transgenerational epigenetic inheritance. *Annual Review of Genetics*, 52, 21–41.

86 Burggren, W.W. (2014) Epigenetics as a source of variation in comparative animal physiology–or–Lamarck is lookin' pretty good these days. *Journal of Experimental Biology*, 217(5), 682–689.

87 Hancock, J.T. (2021) *Cell Signalling*. Oxford University Press, Oxford. ISBN: 9780198859581

88 McCarty, R. (2016) The fight-or-flight response: A cornerstone of stress research. In *Stress: Concepts, Cognition, Emotion, and Behavior* (pp. 33–37). Academic Press, Cambridge, MA (imprint of Elsevier).

89 Organelles are membrane bound sub-compartments inside cells, such as mitochondria and chloroplasts.

90 Ramazi, S. and Zahiri, J. (2021) Post-translational modifications in proteins: Resources, tools and prediction methods. *Database*, 2021.

91 Meenakshi, A.N. and Juvanna, B.I. (2013) Tsunami detection system using unusual animal behavior- A specified approach. *Earth Science*, 2(1), 9–13.

92 Murata, K., Nagasawa, M., Onaka, T., Tsubota, K. and Mogi, K. (2022) Increase of tear volume in dogs after reunion with owners is mediated by oxytocin. *Current Biology*, 32, R855–R873.

93 UK Parliament: https://bills.parliament.uk/bills/2867 (Accessed 05/07/22)

94 Lambert, H., Carder, G. and D'Cruze, N. (2019) Given the cold shoulder: A review of the scientific literature for evidence of reptile sentience. *Animals*, 9(10), 821.

95 Lambert, H., Elwin, A. and D'Cruze, N. (2022) Frog in the well: A review of the scientific literature for evidence of amphibian sentience. *Applied Animal Behaviour Science*, 105559.

96 Lambert, H., Cornish, A., Elwin, A. and D'Cruze, N. (2022) A kettle of fish: A review of the scientific literature for evidence of fish sentience. *Animals*, 12(9), 1182.

97 Lambert, H., Elwin, A. and D'Cruze, N. (2021) Wouldn't hurt a fly? A review of insect cognition and sentience in relation to their use as food and feed. *Applied Animal Behaviour Science*, 243, 105432.

98 Gibbons, M., Sarlak, S. and Chittka, L. (2022) Descending control of nociception in insects? *Proceedings of the Royal Society B*, 289, 20220599. A review of this paper

can also be found at www.iflscience.com/insects-probably-feel-pain-with-big-implications-for-how-we-treat-them-study-says-64349 (Accessed 09/08/22)

99   Rößler, D.C., Kim, K., De Agrò, M., Jordan, A., Galizia, C.G. and Shamble, P.S. (2022) Regularly occurring bouts of retinal movements suggest an REM sleep-like state in jumping spiders. *Proceedings of the National Academy Sciences U.S.A.*, 119, e2204754119

100 Blouin, A.S., Goulet, N., Mills, K., St-Pierre, K. and Harnad, S. (2022) Invertebrate sentience annotated bibliography. https:www.wellbeingintlstudiesrepository.org (Accessed 05/07/22)

101 Shapley, H. (1920) Thermokinetics of *Liometopum apiculatum* Mayr. *Proceedings of the National Academy of Sciences of the United States of America*, 6(4), 204.

102 Martin, J.H. and Bagby, R.M. (1972) Temperature-frequency relationship of the rattlesnake rattle. *Copeia*, 482–485.

103 Guardian: www.theguardian.com/world/2022/aug/17/too-hot-to-chirp-french-heatwave-silences-cicadas-of-provence (Accessed 17/08/22)

104 The BBC: www.bbc.co.uk/programmes/p09x9wmt (Accessed 31/10/21)

105 The Guardian: www.theguardian.com/environment/2020/aug/19/highest-recorded-temperature-ever-death-valley (Accessed 31/10/21)

106 Chemistry Explained: www.chemistryexplained.com/Co-Di/Denaturation.html (Accessed 31/10/21)

107 From Bryn Mawr Classical Review: https://bmcr.brynmawr.edu/2007/2007.09.40/ (Accessed 29/06/22)

108 Ikeya, M. (2004) *Earthquakes and Animals: From Folk Legends to Science.* World Scientific, Singapore. (Note: this is in Italian.)

109 Woith, H., Petersen, G.M., Hainzl, S. and Dahm, T. (2018) Can animals predict earthquakes? *Bulletin of the Seismological Society of America*, 108(3A), 1031–1045.

110 Woith *et al.* (2018).

111 Chum, J., Liu, J.Y., Podolská, K. and Šindelářová, T. (2018) Infrasound in the iono-sphere from earthquakes and typhoons. *Journal of Atmospheric and Solar-Terrestrial Physics*, 171, 72–82.

112 Bedard, A. and Georges, T. (2000) Atmospheric infrasound. *Acoustics Australia*, 28(2), 47–52.

113 The non-protein part of a holistic protein is referred to as a prosthetic group. The classical example is haem in haemoglobin, where the haem is attached to a globin protein (polypeptide).

114 Gould, J.L. (1984) Magnetic field sensitivity in animals. *Annual Review of Physiology*, 46(1), 585–598.

115 Lab Roots: www.labroots.com/trending/plants-and-animals/6835/link-northern-lights-whale-beachings (Accessed 06/01/22)

116 Kirschvink, J.L. (2000) Earthquake prediction by animals: Evolution and sensory perception. *Bulletin of the Seismological Society of America,* 90(2), 312–323.

117 Jeffery, C.J. (2009) Moonlighting proteins – an update. *Molecular BioSystems*, 5(4), 345–350.

118 The reference to Keilin was recently pointed out to me by PhD supervisor, Prof. Owen T.G. Jones, from the University of Bristol (he is long retired, having recently celebrated his 88th birthday).

119 Keilin, D. (1925) On cytochrome, a respiratory pigment, common to animals, yeast, and higher plants. *Proceedings of the Royal Society of London. Series B, Containing Papers of a Biological Character,* 98(690), 312–339.

120 A great book on mitochondria is: Lane, N. (2018) *Power, Sex, Suicide: Mitochondria and the Meaning of Life*. 2nd ed. Oxford University Press, New York. ISBN: 9780198831907.

121 Cytochrome *c* is an instrumental part of the electron transport chain (as mentioned in Chapter 2). The cytochrome shuttles electrons from Complex III to Complex IV. The cytochrome is also partly responsible for giving mitochondria their colour, as oxidised cytochrome *c* is a rich red colour.

122 Bushnell, G.W., Louie, G.V. and Brayer, G.D. (1990) High-resolution three-dimensional structure of horse heart cytochrome c. *Journal of Molecular Biology*, 214(2), 585–595.

123 Jiang, X. and Wang, X., 2004. Cytochrome C-mediated apoptosis. *Annual Review of Biochemistry*, 73(1), 87–106.

124 Apoptosis is named after "falling of leaves" in Greek.

125 Hancock, J.T. (2017) Harnessing evolutionary toxins for signaling: Reactive oxygen species, nitric oxide and hydrogen sulfide in plant cell regulation. *Frontiers in Plant Science*, 8, 189.

126 The Guardian: www.theguardian.com/world/2021/nov/21/volcanic-italian-island-evacuation-increased-activity-vulcano (Accessed 22/11/21)

127 Collins Dictionary: www.collinsdictionary.com/dictionary/english/old-wives-tale (Accessed 02/11/21)

128 Sadewasser, J. (1976) *The Reliability of Selected Weather Beliefs* (Masters theses & Specialist Projects). Paper 2817. WKU: https://digitalcommons.wku.edu/theses/2817 (Accessed 02/11/21)

129 Ewer (1952).

130 Books, and then films, about a wizard, written by J.K. Rowling. Wizarding World: www.wizardingworld.com/ (Accessed 06/01/22)

131 National Geographic: www.nationalgeographic.com/animals/reptiles/facts/komodo-dragon (Accessed 06/01/22)

132 National Geographic: www.nationalgeographic.com/animals/article/dragons-real-komodo-flying-sea-slugs (Accessed 26/02/22)

133 Britannica: www.britannica.com/biography/Zhang-Heng (Accessed 15/11/21)

134 Garstang, M. and Kelley, M.C. (2017) Understanding animal detection of precursor earthquake sounds. *Animals*, 7, 66.

135 NASA, Speed of light: www.grc.nasa.gov/WWW/k-12/Numbers/Math/Mathematical_Thinking/how_fast_is_the_speed.htm (Accessed 24/03/22)

136 NASA, Speed of sound: www.grc.nasa.gov/www/k-12/airplane/sound.html (Accessed 24/03/22)

137 Tributsch, H. (2013) Bio-mimetics of disaster anticipation – learning experience and key-challenges. *Animals*, 3, 274–299.

138 Horsetalk: www.horsetalk.co.nz/2017/09/07/total-eclipse-unsettling-wild-horses/ (Accessed 19/11/21)

139 Greater Good Science Center: https://greatergood.berkeley.edu/article/item/how_modern_life_became_disconnected_from_nature (Accessed 29/06/22)

140 United Nations: www.un.org/en/global-issues/climate-change (Accessed 29/06/22)

141 Shapley (1920).

142 Wehr, T.A. and Helfrich-Förster, C. (2021) Longitudinal observations call into question the scientific consensus that humans are unaffected by lunar cycles. *BioEssays*, 2100054.

143 Scientific American: www.scientificamerican.com/article/can-we-predict-earthquakes-at-all1/ (Accessed 22/11/21)

144 Cleveland Clinic: https://health.clevelandclinic.org/yes-joints-can-predict-weather/ (Accessed 22/11/21)

145 The BBC: www.bbc.co.uk/news/blogs-news-from-elsewhere-33362592 (Accessed 22/11/21)

146 Project Gutenberg: www.gutenberg.org/ebooks/50944 (Accessed 29/06/22)

147 Davy's character called Poietes, probably William Wordsworth. Lamont-Brown, R. (2004) *Humphry Davy: Life Beyond the Lamp*. Sutton Publishing, Stroud, Gloucestershire. ISBN: 9780750932318

148 Davy's character called Ornither, probably an intelligent sportsman and naturalist. The character called Physicus was probably Davy himself. Lamont-Brown (2004).

Appendix

References in full are noted in text and footnotes. Behaviours should be contextualised considering the location where such observations were made (see Chapter 6).

TABLE 1 Examples of Animals Which Can Be Used to Measure or Predict the Environment.

ENVIRONMENT	ANIMAL	BEHAVIOUR NOTED	NOTED BY	COMMENT
Measuring the temperature	Ants	Run speed	Shapley (1920; 1924)	Changes in metabolic rates
	Crickets	Chirp rate	Dolbear (1897) Brooks (1881) W.G.B.[1] (before Brooks) Farmer's Almanac	Only relevant to cold-blooded animals Limited range of temperatures
	Rattle snakes	Rattle rate of tail	Klauber (1940) Chadwick and Rahn (1954) Martin and Bagby (1972)	
	Plant: Rhododendron	Leaf curl	Almanac	Can indicate near-freezing temperatures
Weather				
Predicts rain or storms	Ants	Building walls	Farmer's Almanac	
	Ants	Collect food and then disappear	Enock	Animals preparing for the worst
	Ants	In unusual places, such as ceilings	Enock	Animals preparing for the worst
	Ants	Running back to their nest	"Uncle Offa"	Quoting *The New Book of Knowledge*
	Army ants	Presence of	Radeny *et al.*	Predicts rain
	Bees	Short flight	Farmer's Almanac	
	Bees	Will not leave hive	"Uncle Offa"	Will not swarm before a storm

Centipedes	In large numbers	Theophrastus	
Fleas	Biting	Ewer	
Fleas, gnats and flies	Biting	"Uncle Offa"	Quoting The New Book of Knowledge
Gnats	Height of flight	Ewer	Flying high indicates good weather, with the reverse true. Also a commonly known idea
Black butterflies	Movement of from the south	Radeny et al.	Predicts rain
Flying ants	Occurrence of	Radeny et al.	Enough rain to plant crops
Leaf-cutting ants	Forging more	Sujimoto et al.	Appeared to be sensitive barometric pressure change
Spiders	Length of threads in web	"Uncle Offa"	Long equals fine, short means rain
Spiders	Running around	Risiro et al.	Imminent rain
Leeches	Climb out of water and up the side of bottles	Ewer Used by Merryweather	As used in the Tempest Prognosticator Used commonly by others
Frogs	Croaking during the day	Radeny et al.	Onset of rain
Frogs	Croaking	Enock, Wallisch, "Uncle Offa"	Rain on the way

(Continued)

TABLE 1 (*Continued*)

ENVIRONMENT	ANIMAL	BEHAVIOUR NOTED	NOTED BY	COMMENT
	Toads	Running to their home	"Uncle Offa"	Quoting *The New Book of Knowledge*
	Shark	Extracted liver oil	Ewer	Used in Bermuda
	Fish	Disappeared from reef	Stevenson	On building of the Bell Rock Lighthouse
	Fish (Rāwaru)	Stones are found in the stomach	Māori	Bad weather is coming
	Fish and marine mammals	Swimming upriver	Wallisch	
	Urchin		Bartholomaeus Anglicus	Winds will come
	Sea urchin	Adhering to rock	Wallisch	
	Birds		Observed by Shackleton	Also useful for predicting breakup of the ice
	Birds	Perching in trees	Clark-Platts in *The Cove*	This is a fiction book
	Birds		Stevenson	On building of the Bell Rock Lighthouse
	Swallows	Striking surface of water with bellies	Theophrastus	
	Birds	Fighting for food	Theophrastus	Referred to as a rain crow
	Cuckoo bird	Heard singing	Enock	Rain on the way
	Peacock	Cry	"Uncle Offa"	Quoting *The New Book of Knowledge*

Yellow-billed cuckoo		Georgia, Department of Natural Resources	Referred to as a rain crow
Cocks, hens or hawks	Searching for lice	Ewer	
Kingfisher	Appearance	Enock	Birds sound suggested to sound like rain
Haya birds	Singing and flying	Risiro et al.	Rain on the way
Storks	Singing	Risiro et al.	Rain on the way
Birds	Singing	Radeny et al.	Onset of rain
Swallows	Flying at low altitude	Risiro et al.	Imminent rain
Swallows	Fly low across water	"Uncle Offa"	Quoting The New Book of Knowledge
Ducks or chickens	Stretching their wings and playing in the dust	Radeny et al.	Especially short rains
Ducks	Flutter their wings	"Uncle Offa"	Quoting The New Book of Knowledge
Swallows and swans	Large flocks moving south to north	Radeny et al.	
Falcon (Kārearea)	Screams on a fine day	Māori	Rain the next day, and the opposite applies
Swamp hen (Pūkeko)	Heads for high ground	Māori	
Native parrot (Kākā)	Seen twisting or squawking above the trees	Māori	

(Continued)

TABLE 1 (Continued)

ENVIRONMENT	ANIMAL	BEHAVIOUR NOTED	NOTED BY	COMMENT
	Birds	Flying inland	Wallisch	Also a common saying
	Gulls	Flying inland	Noted by Humphry Davy	Said to "pull in the rain"
	Fork-tailed Swift, Barn Swallow	Presence of	Prober et al.	
	Robin	Sings in bush	"Uncle Offa"	
	Magpies	Number seen	Davy	Can also suggest good weather, depending on number seen
	Doves and pigeons	Arrive in the evening late	"Uncle Offa"	Quoting The New Book of Knowledge
	Sea snakes	Disappear	Liu et al.	
	Tortoise	Feeding	White	
	Seal	Making a loud sound in a harbour while holding an octopus	Theophrastus	
	Hedgehogs, badgers and field mice	Sealing burrows	Ewer	May give indication of direction of wind
	Hare	Seen near harbour	"Uncle Offa"	Storm will follow
	Hedgehog	Blocks burrow	Theophrastus	Violent wind on the way. Which burrow is blocked gives an indication of direction
	Cattle	Fighting for food	Theophrastus	

Animal	Behavior	Source	Notes
Cattle (cows)	Lying down	Ewer	Also in common myths
Cattle (cows)	Lying on their right side, licking their feet, sniffing the air, acting playfully or having lower milk	Ewer	A sign of a storm coming
Cattle	Bellow	"Uncle Offa"	Quoting *The New Book of Knowledge*
Ox	Lick themselves	"Uncle Offa"	Quoting *The New Book of Knowledge*
Cat	Stops frolicking	Swift	
Cat	Play	"Uncle Offa"	
Cat	Running in circles	Clare	Storm approaching
Dog	Eating grass	Wallisch, Ewer	Rain approaching Common myth
Deer and elk	Stampeding	Wallisch	
Sheep	Bleat and play	"Uncle Offa"	Quoting *The New Book of Knowledge*
Swine (pig)	Run with straw on their snout	"Uncle Offa"	Quoting *The New Book of Knowledge*
Mules	Laying back their ears	Wallisch	
Colts (young male horse)	Lying on their backs	"Uncle Offa"	Quoting *The New Book of Knowledge*

(Continued)

217

TABLE 1 (Continued)

ENVIRONMENT	ANIMAL	BEHAVIOUR NOTED	NOTED BY	COMMENT
	Donkeys		Navas González et al.	Bad weather coming
Predicts snow	Hornet	Nest height	Farmer's Almanac	
	Birds	Feeding more	Wallisch	
	Turkeys (Meleagris sp.)	Perching in trees	Wallisch	
	Cats	Facing from fire	Wallisch	
	Hornets	Nesting higher	Wallisch	
Predicts dry weather	Cicadas	Being heard	Farmer's Almanac	Also indicates a coming frost
	Bees	Flight of long distance	Farmer's Almanac	Warm and bright days forecast
	Beetles	Fly in evening	"Uncle Offa"	
	Spiders	Crawling out during rain	Jenner	Rain will be short-lived
	Swallows	Swallows high – staying dry		Common ditty
	Robin	Sings in barn	"Uncle Offa"	Weather will be warm
	Cat	Washes face	Jenner	Fine and clear weather
	Goats	Grazing on high land	Wallisch	Fine weather on the way
	Birds	Singing	Wallisch	
	Goats	Moving to high land	Wallisch	
	Dolphins	Splashing	Wallisch	

Seasons

Season	Animal	Behaviour	Source	Prediction
Winter	Ant	Hill height	Farmer's Almanac	Predicts as early winter
	Crickets	Arrive on hearth	Farmer's Almanac	Long winter ahead
	Sheep and goats	Second breeding season	Theophrastus	
Early spring	Flying squirrel	Calls in the middle of winter	Wallisch	
Rainy season	Crickets	Many on the ground	Enock	Note, this predicts a poor rainy season
	Jerrymanglums/sun spiders	Many around	Enock	Wet season predicted
	Millipedes	Presence of	Risiro et al.	
	Ants	Sealing nests or hunting for food	Risiro et al.	
	Bees and locusts	Large numbers of	Radeny et al.	Will be long rains
	Winged African termite (Coptotermes formosanus Shiraki)	Swarms leave their nests	Okonya and Kroschel	
	Frogs	Presence of	Risiro et al.	
	Dendera birds	Heard singing	Enock	A good rainy season
	Storks	Flying at a very high altitude	Enock	Good season, i.e. good rains

(Continued)

TABLE 1 *(Continued)*

ENVIRONMENT	ANIMAL	BEHAVIOUR NOTED	NOTED BY	COMMENT
	Cuckoo birds (*Cuculiformes: Cuculidae*)	Start to call	Okonya and Kroschel	
	Bittern (Matuku-hūrepo)	Squawking of	Māori	Suggests a season of floods
	Heron (Kōtuku)	Presence of many in the summer	Māori	Gales and heavy snow will follow
	Ground squirrel	Digging of holes	Radeny *et al.*	Indicative of a normal rainy season
	Goats	Breeding	Risiro *et al.*	
	Monkeys, leopards, antelope and baboons	Moving into village	Radeny *et al.*	Season of good rains
Dry season	Bush crickets (*Ruspolia baileyi* Otte)	Appearance of	Okonya and Kroschel	
	Migratory birds such as cattle egrets (*Bubulcus ibis* Linnaeus)	Appearance and movement of	Okonya and Kroschel	
	Bateleur eagle (*Terathopius ecaudatus* Lesson)	Calling by	Okonya and Kroschel	
Warm season	Shining cuckoos (Pīpīwharauroa) or godwits (Kūaka)	Arrival of	Māori	

	Animal	Behaviour	Reference	Notes
	Isabella tiger moth (woolly worm)	Banding pattern of caterpillar	Curran, Sloane and others	Wider banding indicates mild season – opposite also noted
	Groundhog	Leaving borrow	Wallisch and others	Warm weather: opposite also noted; Common myth
Hot season	Insects	Singing of	Risiro *et al.*	
	Reptiles	Large number of	Risiro *et al.*	
	Rock rabbit	Crying in the morning and evening	Risiro *et al.*	
Good season	Insects	Presence on Albizia trees with water dripping from them	Radeny *et al.*	
	Stork, denderas, swallows	Appearance	Enock	Also predicts rain
	Game animals	Giving birth in large numbers	Enock	And the opposite also applies
Drought	Waterfowls	Breeding on the ground and in lower patches on flood plains	Enock	No need to protect themselves from bad weather
	Hawks	Perching in tree	Enock	Ending of drought
	Animals	High birth rate of males	Radeny *et al.*	Arrival of a drought season
Earthquakes	Centipedes	Move to safety	Ancient Greek	Proceeds before earthquake
	Bees	Swarming	Bhargava *et al.* & Tributsch	

(Continued)

TABLE 1 (Continued)

ENVIRONMENT	ANIMAL	BEHAVIOUR NOTED	NOTED BY	COMMENT
	Toads	Stopped spawning	Grant and Halliday	5 days before
	Snakes	Move to safety	Ancient Greek	Proceeds before earthquake
	Snakes	Come out of hibernation	Bhargava et al. "Uncle Offa"	
	Snakes	Appeared "frozen" to the roads	Smith	
	Catfish	Jumping in a pond	Bhargava et al.	One day before
	Fish	Swimming in circles and jumping from the water	Tributsch	
	Seagulls[2]	Fly inland	Internet source	1-2 days before
	Birds (several species)	Took flight	Yosef	Small birds not affected, timescale – minutes
	Chickens	Refusing to enter their coops	Smith	
	Chickens	Flying into the trees/refusing to go in pens	Tributsch	
	Chickens	Not roost and stay outdoors	"Uncle Offa"	
	Geese	Taking flight	Smith	
	Ducks and geese	Avoiding water	Tributsch	

Pigeons	Avoiding their lofts	Tributsch	
Birds	Screaming	Bhargava et al. & Tributsch	
Rats, weasels	Move to safety	Ancient Greek	Proceeds before earthquake
Rats	Insulin metabolism altered	Chen et al.	
Rats	Appearance of	Bhargava et al.	
Rats	Dazed	Smith	
Rats	Running out of their holes	Bhargava et al. & Tributsch	
Mice	Circadian rhythms disrupted	Yokoi et al. Li et al.	
Mice	Behaving strangely	Tributsch	Before major earthquake in China
Rabbits	Refusing to eat and trying to leave their pens	Tributsch	
Elephants	Sense movement/ sound	Listed by BBC	Altered behaviour prior to event
Dogs and cats	Unusual behaviour	Yamauchi et al.	Day before
Companion animals (pets)	Disappearing	Reported in San Francisco	Little evidence of this being true
Dogs	Barking and removing their puppies from the house	Tributsch	

(Continued)

TABLE 1 (*Continued*)

ENVIRONMENT	ANIMAL	BEHAVIOUR NOTED	NOTED BY	COMMENT
	Cats	Carry kittens outdoors	"Uncle Offa"	
	Cats	Restlessness	Tributsch	
	Dogs	Barking	Bhargava et al. & Tributsch	
	Cats	Jumping out of windows	Bhargava et al. & Tributsch	
	Cows and horses,	Appearing restless	Smith	
	Cows	Decreased milk production	Yamauchi et al.	During week before
	Cows, dogs, and sheep	Bio-logging survey	Wikelski et al.	Animals indoors better than those outside: up to 20-hour prediction
	Horses	Run wild, kick and rear	"Uncle Offa"	
	Deer	Leave the woods or forest	"Uncle Offa"	
	Bears	Come out of hibernation	"Uncle Offa"	
	Pigs	Restless	"Uncle Offa"	
Tsunami	Birds	Took to the air	Meenakshi and Juvanna	
	Birds	Calling that "something was wrong"	Tiwari and Tiwari	
	Sea turtles	Observations	Jain et al.	Used in an algorithm
	Elephants	Trumpeted and ran to higher ground	Meenakshi and Juvanna	

Volcanoes	Snakes	Incidence of bites	Kurniawan et al.	
	Goats	Movement on mountain	Icarus Global Monitoring	6 hours warning
Events in the sky				
Lunar events	Insects	Lunar effects	Zimecki	
	Amphibians	More spawning at full moon	Grant et al.	
	Fish	Lunar effects	Zimecki	
	Fish	Reproduction affected by moon	Takemura et al.	May be hormonal effects
	Birds	Lunar effects	Zimecki	
	Rats/mice	Lunar effects	Zimecki	
	Deer	Affected by moonrise and moon set	Sullivan et al.	Well known by hunters / No prediction of events
	Donkeys	Affected by lunar cycles	Navas González et al.	Probably a stress response / No prediction
	Sheep	Lunar effects	Zimecki	

(Continued)

TABLE 1 (Continued)

ENVIRONMENT	ANIMAL	BEHAVIOUR NOTED	NOTED BY	COMMENT
	Cows	Birth rates affected by lunar cycle	Yonezawa et al.	
	Humans	Some effects reported	Bevington	
Solar eclipse	Insects/dung beetles (Paragymnopleurus planus)	Unusual behaviour	Wiantoro	No prediction
	Amphibians	Unusual behaviour	Wiantoro	No prediction
	Black Flying Fox (Pteropus alecto)	Became still	Wiantoro	No prediction
	Dian's Tarsier (Tarsius dentatus)	No response	Wiantoro	No prediction
	Pig (Sus sp.)	Showed abnormal behaviour	Wiantoro	No prediction
	Heck's macaque (Macaca hecki)	Gave out a call	Wiantoro	No prediction
	Maleo (Macrocephalon maleo)	Behaved as though night	Wiantoro	No prediction
	Chimpanzees (Pan troglodytes)	Pointed to sky	Branch and Gust	No prediction
	Baboons (Papio hamadryas)	Unusual behaviour	Gil-Burmann Beltrami	No prediction

	Rats	More active	Sethi	No prediction
	Rabbits	Less active	Sethi	No prediction
	Marine animals	Unusual behaviour	Kumar and Rengaiyan	No prediction
	Horses	Unusual behaviour	Maki	No prediction
Northern Lights	Dogs	Look to the lights	Internet source	No prediction
	Whales	Become beached	Granger	No prediction

1 The identity of W.G.B. has remained unknown.
2 As indicated elsewhere, seagulls do not exist as a species, but it is a common catch-all term for gulls.

TABLE 2 Examples of Environmental Factors Which May Be Sensed for Prediction of Change.

ENVIRONMENTAL FACTOR	EXAMPLES OF ANIMALS	MECHANISM OF SENSING/COMMENTS
Light		
Visible (human) wavelengths	Human	Limited to approximately 400 nm to 700 nm Eyes – retinal in the rods and cones react with photons
Wavelengths outside human perception	Birds	Near ultraviolet (320 nm–400 nm)
	Snakes	Near infrared
Sound		
Audible	Humans	20–20,000 Hz, but higher pitch perception drops with age. Ears – mechanical movement of hairs
	Elephants	Down to 17 Hz or lower
	Bats	5 Hz to 200 Hz – used for echolocation
	Gulls	May sense pulse of sound before a storm?
Infrasound	Birds	Used for navigation. Pigeons sensitive down to 0.05 Hz. Perhaps used in seismic sensing.
	Pigeons and guinea fowl (*Numida meleagris*)	Use infrasound-sensitive neurons which are situated in the midbrain
	Capercaillie (*Tetrao urogallus*)	Can produce infrasound
	Peacocks	Produce and respond to infrasound during their mating displays
	Elephants	Can produce infrasound
	Elephants	May use infrasound for seismic sensing

Chemicals

Smells	Primates	Interacting with receptors in the nose
	Birds	Some may smell their way home for navigation
	Fish	Some fish "smell" the water as navigation, e.g. salmon
Pheromones	Many insect species	Many animals (e.g. bees) release and sense small chemicals

Barometric pressure

	Ectoparasitoid (*Mallophora ruficauda*)	Changes in host-seeking behaviour
	Leaf-cutter ants (*Atta sexdens*)	Nest leaving and leaf cutting altered
	Curcurbit beetle (*Diabrotica speciosa*)	Altered behaviour: hair-like receptors in the cuticle
	True armyworm moth (*Pseudaletia unipuncta*)	Altered behaviour: hair-like receptors in the cuticle
	Potato aphid (*Macrosiphum euphorbiae*)	Altered behaviour: hair-like receptors in the cuticle
	Parasitic wasp (*Leptopilina heterotoma*)	Altered egg laying
	Frogs	Croaking altered
	Birds – in general	Many animals are known to be sensitive to barometric pressure changes
	White-crowned sparrow (*Zonotrichia leucophrys*)	Sense and respond to declining barometric pressure

(Continued)

TABLE 2 (Continued)

ENVIRONMENTAL FACTOR	EXAMPLES OF ANIMALS	MECHANISM OF SENSING/COMMENTS
	Ducks	Capable of discriminating changes as small as 0.4 p.s.i
	Pigeons	50% threshold of effect was approximately 10 mm H_2O
	Sharks (*Carcharhinus limbatus*)	Swim deep as pressure changes
	Bat (*Pipistrellus subflavus*)	May be indicating good feeding time
	Sea snakes (*Laticauda* spp	Go deep to avoid storms
	Mice	Pain responses. Pressure dropped 40 hPa for 50 minutes, and gene expression altered.
	Rat	Alterations seen in swim test
	Squirrel	Activity increases as barometric pressure increases
Vibrations		
	Amphibians	Known to sense ground movements
	Birds	Probably sense vibrations
	Vertebrates	Elephants are a good example of sensing ground movements
Ground tilting		
	Some animals	May be sensitive to ground angles
Humidity		
May not be useful in naturally humid climates, and not at all in aquatic environments	Mammals	Hair is sensitive to humidity – basis of some hygrometers
	Insects	May be sensitive
	Spiders	May be sensitive
	Plants	Some plant foliage is sensitive to changes

Magnetism/electrical

Geomagnetism	Bacteria	May contain magnetic material, e.g. Fe_3O_4
	Insects – honeybees	May have similar magnetic sensors
	Fish – tuna	May have similar magnetic sensors
	Fish – salmon	May use geomagnetism for navigation
	Birds – (pigeons)	Used for navigation. Not all birds affected.
	Whales	Used for navigation/beaching of whales
	Rodents	Altered circadian rhythms
	General	Some animals may have cryptochromes – may be important for nocturnal animals
Electrophonics	General	Generate a range of sounds with frequencies ranging between 20 Hz and 20 kHz, which can then be sensed by many animals
Changes in atmospheric ions	Toads	Stopped spawning: may be sensing sound

Gravity

	Amphibians	Spawning may be affected
	Insects	Suggested to be sensitive
	Fish	Suggested to be sensitive
	Birds	Suggested to be sensitive
	Rats	Suggested to be sensitive
	Mice	Suggested to be sensitive
	Donkeys	Thought to be sensitive
	Sheep	Suggested to be sensitive

(Continued)

TABLE 2 (Continued)

ENVIRONMENTAL FACTOR	EXAMPLES OF ANIMALS	MECHANISM OF SENSING/COMMENTS
Changes in water pH		
	Aquatic animals	May be sensitive to changes in water acidity
Temperature		
	Many animals	Sense and alter activity as temperature changes
	Many animals	Some animals contain antifreeze systems so can go to low temperatures, e.g., fish
	Ants, snakes, crickets	Alter activity and can be used as a proxy thermometer
	Bacteria	Some organisms are thermophiles and exist at relatively high temperatures

Note: Many of these systems are suggested or speculative, with little robust experimental evidence. As well as the environment factors listed next, many animals will be sensitive to wind (or water currents) and precipitation, such as rain or snow. These factors are not listed, but should not be ignored in observing how animals survive in their environments.

TABLE 3 Some of the Technologies Which Can Be Used to Measure Environment.

CONDITION MEASURED	TECHNOLOGY	EXAMPLES OF TYPES FOUND	COMMENTS
Temperature	Thermometer	Thermoscope	Early version by Galileo Galilei
		Mercury-based	A version produced by Fahrenheit in 1714
		Alcohol-based	A version produced by Fahrenheit in 1709
		Digital	Based on Seebeck's work
		For use in ear	Benzinger's work
		Electronic Cricket	**Based on Dolbear's Law**
Air pressure	Barometer	Mercury	Torricellian tube was the basis for those that followed
		Aneroid	Sealed metal box: invented by Lucien Vidi
		Digital	More accurate and easier to carry
Wind speed	Anemometer	Cups rotating horizontally	Can have variable number of cups Oldest version perhaps Leon Battista Alberti in 1450 Modern version credited to John Robinson, in 1846
		Sonic anemometer	Invented by Andreas Pflitsch in 1994
	Windsocks	Usually a hollow tube of material	Can give an indication of wind strength but not quantifiable Also give wind direction
Wind Direction	Weathervane	A tail ensures that the vane points into the wind	Otherwise known as a windvane Invented by Andronicus, in 48 BC

(Continued)

TABLE 3 (*Continued*)

CONDITION MEASURED	TECHNOLOGY	EXAMPLES OF TYPES FOUND	COMMENTS
	Windsocks	Usually a hollow tube of material	Often seen at airports or on high bridges
	Anemometer	Ultrasound variety	Ultrasound anemometers can measure speed and direction of the wind
Rain	Rain gauge	Simple catchment for water	Often measured manually
		Weight gauge	Can measure weight of fallen precipitation
		Tipping bucket	A pivot/scale system
		Optical	Laser diode and detector
		Acoustic	Relies on sound signature of fallen drops
		Modern home weather station	Home weather monitors also measure precipitation
Snow	Measuring stick		Often snow gauge sticks are used along mountain roads or other places where snow may disrupt travel
	Snow gauge		
Humidity	Hygrometer		Originally invented by Leonardo da Vinci, although used in Shang Dynasty (1556–1046 BC)
		Hair	Curls as it absorbs water, e.g. as used by Horace-Bénédict de Saussure
		Any material that moves with water	Absorbs water and moves, e.g. Hooke used the husk of an oat grain

		Dew-point	Condensation of water
		Two thermometer/psychrometer	One thermometer is kept wet, and hence they record difference temperatures, depending on evaporation
Weather modelling	Supercomputers		Based on semiconductors
			e.g. Roshydromet technology
Weather in general	Home weather stations	Digital	Can output to internet sites for further data analysis
		Several now commercially available	
	Tempest Prognosticator	**Leeches**	**As invented by Merryweather, said to predict bad weather**
	Shark oil barometer	**Oil extracted and placed in bottle**	**Said to predict bad weather Common in Bermuda**
	Satellites		Send images/data back to earth
			Can also see other phenomena such as ash volcanic ash clouds
Weather networks	Crowd-based internet resources	Taking data from "home" stations	e.g. Automatic WEather KArten System (AWEKAS), Weather-cloud, and Weather Underground
Earth tremors	Seismometer	Falling ball	Invented by Zhang Heng (78–139)
		Pendulum	e.g. Scotland in 1840, also developed by Filippo Cecchi (1822–1887)
		Mercury	Invented by Luigi Palmieri in 1855
		Strain-based	Hugo Benioff (1899–1968)
	Monitoring networks	Gather information for warnings	e.g. Northwest Pacific Tsunami Advisory Center (NWPTAC), Earthquake Warning California, ShakeAlert

(Continued)

TABLE 3   *(Continued)*

CONDITION MEASURED	TECHNOLOGY	EXAMPLES OF TYPES FOUND	COMMENTS
Stars and space	Telescopes	Light	Early version invented by Galileo, but versions also by Hans Lipperhey and others
		Light	Improvements by Newton etc.
		Radio	Based on the work of Karl Guthe Jansky and Grote Reber
		Infrared	e.g. Herschel Space Observatory
	Space telescopes	e.g. Hubble & James Webb	Unhindered by atmosphere
Northern Lights	Phone apps		Useful for prediction
	Webcams		Can be mounted to 'watch' for lights
	Internet sites		Several predict where the lights will be able to be seen
	Viewed	Eyes or cameras	Can be seen with the naked eye but enhanced by cameras
Lunar cycles	Known trajectories	Mathematics	Known lunar (and planetary) movements can be used to predict timings and height of tides
Eclipses (solar and lunar)	Known trajectories	Mathematics	Known movements of Earth, moon and sun can be used to calculate exact timings

Note: The Southern Lights (*Aurora australis*) are not listed here as they are less likely to be seen from land. The technologies based on animals are highlighted in bold.

Index

Milton Keynes UK
Ingram Content Group UK Ltd.
UKHW020049091123
432234UK00018B/118